Geography of the Physical Environment

The *Geography of the Physical Environment* book series provides a platform for scientific contributions in the field of Physical Geography and its subdisciplines. It publishes a broad portfolio of scientific books covering case studies, theoretical and applied approaches as well as novel developments and techniques in the field. The scope is not limited to a certain spatial scale and can cover local and regional to continental and global facets. Books with strong regional focus should be well illustrated including significant maps and meaningful figures to be potentially used as field guides and standard references for the respective area.

The series appeals to scientists and students in the field of geography as well as regional scientists, landscape planners, policy makers, and everyone interested in wide-ranging aspects of modern Physical Geography. Peer-reviewed research monographs, edited volumes, advance and undergraduate level textbooks, and conference proceedings covering the major topics in Physical Geography are included in the series. Submissions to the Book Series are also invited on the theme 'The Physical Geography of…', with a relevant subtitle of the author's/editor's choice.

More information about this series at http://www.springer.com/series/15117

Kalyan Rudra

Rivers of the Ganga-Brahmaputra-Meghna Delta

A Fluvial Account of Bengal

Kalyan Rudra
Environment Department
West Bengal Pollution Control Board
Kolkata, West Bengal
India

ISSN 2366-8865   ISSN 2366-8873 (electronic)
Geography of the Physical Environment
ISBN 978-3-319-76543-3   ISBN 978-3-319-76544-0 (eBook)
https://doi.org/10.1007/978-3-319-76544-0

Library of Congress Control Number: 2018934383

© Springer International Publishing AG, part of Springer Nature 2018
This work is subject to copyright. All rights are reserved by the Publisher, whether the whole or part of the material is concerned, specifically the rights of translation, reprinting, reuse of illustrations, recitation, broadcasting, reproduction on microfilms or in any other physical way, and transmission or information storage and retrieval, electronic adaptation, computer software, or by similar or dissimilar methodology now known or hereafter developed.
The use of general descriptive names, registered names, trademarks, service marks, etc. in this publication does not imply, even in the absence of a specific statement, that such names are exempt from the relevant protective laws and regulations and therefore free for general use.
The publisher, the authors and the editors are safe to assume that the advice and information in this book are believed to be true and accurate at the date of publication. Neither the publisher nor the authors or the editors give a warranty, express or implied, with respect to the material contained herein or for any errors or omissions that may have been made. The publisher remains neutral with regard to jurisdictional claims in published maps and institutional affiliations.

Cover image by Sonja Weber, München

Printed on acid-free paper

This Springer imprint is published by the registered company Springer International Publishing AG part of Springer Nature
The registered company address is: Gewerbestrasse 11, 6330 Cham, Switzerland

Dedicated in memory of my parents who taught me basic principles of the Nature during my childhood.

Foreword

The Ganga–Brahmaputra–Meghna (GBM) delta largely coincides with the political boundary of undivided Bengal. Three mighty river systems have constituted an intricate drainage network which drains Bengal. The river network in this geologically new land has changed dramatically since the late eighteenth century when first systematic survey of Bengal was conducted by James Rennell (1780) and 'A Bengal Atlas' was published. Since then many rivers have either decayed or changed their courses. The life and culture of Bengal are intertwined with its river system. The fertile land, copious rainfall and bright sunlight played pivotal role in making Bengal agriculturally prosperous.

A holistic research on river system of the GBM delta has never been undertaken since area was politically divided in 1947. Eminent scholars worked mostly on Bangladesh, and some outstanding articles have been published. The western part of the delta was not given proper heed. The fragmented research on the GBM system does not offer comprehensive understanding. This book is the only publication which deals with river system of undivided Bengal after a long gap of about eight decades.

Kolkata, India   Kalyan Rudra

Preface

The Ganga and the Brahmaputra enter Bengal[1] through Rajmahal–Meghalayas gap and constitute an intricate drainage network draining vast alluvial terrain which is one of the most dynamic hydro-geomorphological units of the world. The Barak and Surma are two important headwaters constituting the Meghna system which drains Sylhet basin of Bangladesh. The Ganga–Brahmaputra–Meghna delta is a part of larger tectonic terrain which is known as the Bengal basin. The mighty fluvial system played a key role to fill up basin and now carries 1375 BCM water with more than one billion tons of suspended sediment loads per annum. In this vast alluvial tract, the rivers have the tendency to move to and fro, like that of the pendulum of a wall clock, within the meander belts. The other form of dynamism is the avulsion which may happen during a devastating flood, or the process may be slow, covering a protracted period. The eastward avulsion of the Teesta in 1787, the southward detour of Damodar in mid-seventeenth century, the westward migration of the Brahmaputra (known as Jamuna in Bangladesh) in 1830, the eastward flight of the water of the Ganga through the Padma (name of the main distributary of the Ganga in Bangladesh) and recent westward diversion of the waters of the Mahananda through the Fulohar which takes off at Bagdob are five examples of avulsions. Many palaeo-channels of the delta remind us the dynamics of the river during the historical period.

The flowing waters in rivers of Bengal during the late monsoon exceed the critical limit and submerge the adjoining floodplain. The plenty of water, bright sunlight and fertile alluvial tract made Bengal ecologically productive and agriculturally prosperous. But colonial river management plan brought about serious drainage congestion and aggravated decay of many channels, especially which were trapped between earthen embankments (locally called *bund*). The assurance of protection against floods through the building of linear embankment along the bank of rivers ultimately proved to be futile. The frequent breaches in the embankment had been common. The ecological degeneration relating to embankment building was manifold, but the most important was that the farmlands were deprived from the annual deposition of fertilizing silts. The condition deteriorated further when railways and roads were built on high embankments to ensure uninterrupted movement even during the flood. The earth materials for building embankments were

---

[1]Bengal includes West Bengal in India and the Bangladesh together.

borrowed from ditches excavated parallel to the lines of communication, and those borrow pits subsequently became swamps facilitating breeding of malarial mosquito. The spread of malaria followed the paths of railway and highway. Since the railways and highways in both north and south Bengal were transverse to the flow direction of rivers and were built with narrow culverts, there were drainage congestions, expansion of flood contours and decline in agricultural productivity.

The Farakka barrage was built across the Ganga to induce water into the Bhagirathi-Hugli river and improve the navigational status of the port of Kolkata. Since 87 million cubic metres of water was impounded, the Ganga started to change its course and many villages were engulfed by the river. Some other rivers of Bengal were dammed to facilitate irrigation and hydro-power generation during post-independence era. Since the engineers had a lack of understanding about the transmission–distribution loss of water in dam–canal network, the gap between potentially irrigable area and the area actually irrigated gradually widened and there was increasing dependence on groundwater to fill up this gap. The exploitation of groundwater exceeding replenishable limit led to the diminution of base flow and ultimately desiccation of rivers. In this politically divided GBM delta, fifty-four rivers continue to flow cutting across international border and the equitable and rational sharing of transboundary flow has emerged as an issue of hydro-diplomacy. In both India and Bangladesh, there are tendencies to overexploit the rivers to ensure irrigation, especially during the lean months. However, maintaining ecological flow of rivers is critical to the sustenance of biodiversity along with the well-being of the mankind who depend on the rivers. It is important to explore the threshold between the volume of water that may be extracted from the rivers and the flow to be maintained in the rivers to ensure the ecosystem services. Both India and Bangladesh need to find a mutually acceptable policy of river management for the greater common good.

This book discusses all the issues stated in the foregoing paragraphs.

Kolkata, India   Kalyan Rudra

Acknowledgements

I express my deep gratitude to Prof. Debasis Sengupta of Indian Statistical Institute, Kolkata, for his academic support to estimate flow of rivers through a mathematical modelling. Late Prof. Graham Chapman of Lancaster University, UK, had a keen interest in rivers of Bengal, and I was immensely benefited while discussing with him on several issues.

I must also express my grateful thanks to Ms. Anwesha Halder, presently Senior Research Fellow, Department of Geography, University of Calcutta, and Ms. Debasmriti Chaudhuri, Department of Geography, Asutosh College, Kolkata, who worked with me as Research Assistants in West Bengal Pollution Control Board and also offered cartographic assistance along with Dr. Mafizul Haque of Department of Geography, University of Calcutta. Professor Sumita Banerjee of Loreto College, Kolkata, reviewed the manuscript. I am especially thankful to Ms. Gouri Rudra and Sri Debojyoti Datta also for their constant support.

For last three years, I have been working as the Chairman of West Bengal Pollution Control Board. I am indebted to my colleagues for their constant support and intellectual ambience in WBPCB which had been the backdrop to all my work.

Contents

1 Rivers of the Ganga–Brahmaputra–Meghna Delta: An Overview 1
    1.1   The Ganga–Brahmaputra–Meghna (GBM) Delta 3
    1.2   Dynamic River System 4
    1.3   Rivers of Barind Tract 7
    1.4   The Ganga and the Bhagirathi–Hugli 8
    1.5   The Rivers of Rarh Bengal 9
    1.6   The Brahmaputra-Meghna System 9
    1.7   The Tidal Creeks 10
    1.8   Managing the Rivers 10
    1.9   Ecological Flow 13
    References 13

2 Evolution of the Bengal Basin 15
    2.1   Structure of the Basin 17
    2.2   Basin-Fill History 17
    2.3   Late Quaternary Sequences 19
    2.4   The Bengal Basin and the GBM Delta 20
    2.5   Stratigraphy of Delta 23
    2.6   The Changing Coastline 24
    References 24

3 Rivers of the Tarai–Doors and Barind Tract 27
    3.1   Mountains to Plains 29
    3.2   Rivers of the Himalayan Front 30
    3.3   Characteristics of North Bengal Rivers 30
    3.4   The Teesta and Its Tributaries 31
    3.5   The Teesta Fan 34
    3.6   Human Intervention and River Style 36
    3.7   The Jaldhaka System 37
    3.8   The Tributaries 38
    3.9   The Torsa System 40
    3.10   The Mahananda River System 41
    References 47

4	**The Dynamic Ganga**		49
	4.1	The Ganga through West Bengal	51
	4.2	The Dynamic Ganga	55
	4.3	The Changing Course of the Ganga	55
		4.3.1 From Rajmahal to Farakka	55
	4.4	Changing Course Between Farakka and Jalangi	59
	4.5	Mechanism of Changing Course	62
	4.6	Estimated Flow	66
	4.7	The Ganga System in Bangladesh	66
	4.8	Policy Issues	68
	References		70
5	**The Jamuna–Meghna System**		73
	References		76
6	**The Bhagirathi-Hugli River System**		77
	6.1	Antiquity of the Bhagirathi	79
	6.2	The Bhairab–Jalangi and the Mathabhanga–Churni	80
	6.3	The Changing Off-take of Bhagirathi	81
	6.4	Meandering Channels	83
	6.5	Dying Rivers	85
	6.6	The Saraswati River	87
	6.7	The Bidyadhari–Sunti–Noai System	87
	6.8	The Jamuna and the Ichhamati	88
	6.9	The Adi Ganga	89
	6.10	The Hugli Estuary	89
	6.11	Suspended Load	91
	References		92
7	**The Western Tributaries to the Bhagirathi–Hugli River**		95
	7.1	The Bansloi	95
	7.2	The Pagla River	97
	7.3	The Mayurakshi	97
	7.4	The Ajay	98
	7.5	The Damodar	100
	7.6	The Khari–Banka	102
	7.7	The Rupnarayan	102
	7.8	The Kansai or Kangshabati	103
	7.9	The Rasulpur	104
	References		106
8	**The Sundarban**		107
	8.1	Hydro-Geomorphological Characteristics	110
	8.2	The River System	112
	8.3	The Major Channels of the Sundarban	112
	8.4	Early Civilization	115
	8.5	Premature Reclamation	117
	8.6	Protection with Embankment: Myth and Reality	118
	8.7	Impacts of Tropical Cyclones	120
	8.8	Recent Challenges	121

	8.9	Rising Temperatures	121
	8.10	Sea-Level Changes	121
	References		124

9 Flood in the GBM Delta ... 125
- 9.1 Introduction ... 125
- 9.2 Changing Rainfall Pattern ... 126
- 9.3 Location and Topographic Expression ... 127
- 9.4 Major Floods of Bengal ... 128
  - 9.4.1 Flood Management ... 130
- 9.5 Issues of Disaster Management ... 133
  - 9.5.1 Early Warning ... 134
  - 9.5.2 Strengthening Preparedness ... 134
  - 9.5.3 India's National Policy on Floods ... 134
  - 9.5.4 Policy in Bangladesh ... 135
  - 9.5.5 Linking Flood Management with Development ... 135
- References ... 136

10 Management of Rivers in the GBM Delta ... 137
- 10.1 Colonial Period ... 137
- 10.2 Altered Hydrological Regime ... 138
  - 10.2.1 Structural Interventions ... 141
  - 10.2.2 Pre-Mature Land Reclamation ... 142
- 10.3 Post-Colonial Period ... 142
  - 10.3.1 Damodar Valley Corporation ... 143
  - 10.3.2 PMUD Summary and Observations ... 145
  - 10.3.3 Power Generation and Irrigation ... 148
  - 10.3.4 Flood Control ... 149
  - 10.3.5 DVC: An Overview ... 150
  - 10.3.6 Mayurakshi Project ... 152
  - 10.3.7 The Farakka Barrage Project ... 153
- 10.4 The Teesta Barrage Project ... 157
- 10.5 The Teesta Barrage Project in Bangladesh ... 161
- References ... 161

11 Conflicts Over Sharing the Waters of Transboundary Rivers ... 163
- 11.1 Provisions in the International Rules ... 164
- 11.2 The Rivers Crossing Indo-Bangladesh Border ... 164
- 11.3 Sharing the Ganga Water ... 166
- 11.4 Conflict Over the Teesta Water ... 168
  - 11.4.1 The Teesta Basin ... 168
  - 11.4.2 Unrealistic Planning ... 168
  - 11.4.3 Hydro-Politics and Negotiations ... 169
  - 11.4.4 Gap in the Irrigation ... 169
  - 11.4.5 Water for Power Generation ... 170
  - 11.4.6 Maintaining Ecological Flow ... 170
- 11.5 Two Barrages: Myth and Reality ... 171
- References ... 171

12	**The Concept of Ecological Flow**	173
	12.1 River Ecology	174
	12.2 Impact of the Altered Flow Regime	176
	12.3 Defining the Concept of Ecological Flow	177
	12.4 An Engineering-Management Approach	177
	12.5 A Rational Meeting Point	178
	12.5.1 The Indian Scenario	179
	12.5.2 Experience of Bengal	182
	References	184
Index		187

Abbreviations and Units

AD	Anno Domini
BC	Before Christ
BCM	Billion cubic metre
BDPUB	Bangladesh Pani Unnyan Board
BP	Before present
Cumec	Cubic metre per second
Cusec	Cubic feet per second
DVC	Damodar Valley Corporation
EMC	Environmental management classes
EWR	Environmental water requirement
FDC	Flow duration curve
IPCC	Intergovernmental Panel on Climate Change
IUCN	International Union for Conservation of Nature
IWA	International Water Association
m.s.l.	Mean sea level
MAR	Mean annual run-off
MCM	Million cubic metre
MLD	Million litres per day
MW	Megawatt
TJMC	Teesta–Jaldhaka link cabal
TMLC	Teesta–Mahananda link canal
WCD	World Commission on Dams

Rivers of the Ganga–Brahmaputra–Meghna Delta: An Overview

1

Abstract

The Ganga–Brahmaputra–Meghna delta covering an area of about 200,000 km² has emerged by the process of deposition on a shallow tectonic basin which is also known as Bengal due to its cultural and linguistic identity. It is an area of interlacing drainage and fast-changing landscape. Many rivers of Bengal have changed their courses by meander migration and avulsion during last three centuries, and the striking changes have been observed in the coastal tract. While the sea has been encroaching inland along Indian part, fast accretion is recorded along the Meghna estuary in the east. The agricultural prosperity of Bengal is intertwined with its fluvial system. But since many rivers were embanked to control the flood, the sediment loads were trapped in rivers and agricultural fields were deprived from annual replenishing of fertility. The dams and barrages built across the rivers interrupted downstream transfer of sediment load and created longitudinal disconnectivity in rivers. Thus, decay of channels was aggravated due to human interventions. This chapter is an overview of the dynamic river system of the GBM delta in the subsequent chapters.

The GBM delta is located within political boundary of undivided Bengal and truly called the land of rivers. It covers the state of West Bengal (India) and Bangladesh and contains the largest delta of the world. It is geologically described as a basin which has been filled up by the deposition of sediment load. The landscape is fast changing as a result of the very large monsoon discharges and sediment loads of three mighty rivers, the Ganga, the Brahmaputra and the Meghna, and many tributaries which debouch from the Himalaya in the north, the Chotanagpur plateau in the west and Manipur–Tripura hills in the East into a large tectonic unit called Bengal basin (Fig. 1.1). The basin almost coincides with the cultural and linguistic region known as Bengal which was politically divided in 1947 and is now shared by India and Bangladesh. The international boundary was intended to divide the region between Muslim and non-Muslim region, and consequently the line crossed over 54 rivers, creating the scope for potential conflicts over sharing of flowing waters (Bagge 1950).

Bengal is often described as *nadimatrik* (a land nurtured by rivers as mothers) but the rivers have been decaying fast, posing a serious threat to lives, culture and economy. Many rivers have disappeared and many have gone dry (Rudra 2015). The perennial rivers have turned ephemeral. Some major rivers have been continuously changing the geometry of meanders causing erosion of fertile agricultural land and displacement of large numbers of people. The flood contours are expanding. Even the Ganga is in an

Fig. 1.1 Major rivers of GBM delta

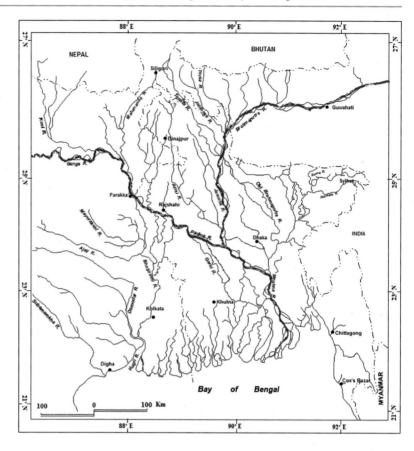

alarming state of decay which means not only dwindling discharge and increasing sedimentation but also deteriorating quality of water. Unfortunately, the Ganga has been used as the most convenient outlet of wastewater and consequently the water is not even fit for bathing at many places (Rudra 2009). There is dramatic lack of understanding at the official level how the river system functions in the Bengal basin. The flawed logic of river management often makes the situation complicated. The holistic river research is impeded since river basins are politically divided, and the flow data are declared as 'classified'. Those data are not available in public domain.

The colonial rulers tried to appreciate the dynamics of river system, and many scholarly reports were published from the late eighteenth century till the first half of the twentieth century. The West Bengal District Gazetteer Department has reprinted some of those reports as a compilation named 'Rivers of Bengal' in four volumes. Bangladesh Water Development Board had divided the country into six hydrological regions and published exhaustive reports in 2011. It is important to appreciate colonial history to have a better understanding of the human relationship with rivers in Bengal. The British rulers were much concerned for the gradual decay of the Bhagirathi–Hugli River and the consequent obstacle to the navigation in and out of the Kolkata (erstwhile Calcutta) port. The colonial research was intended to explore the possible ways of improving navigation and finding best possible channels for sailing. They also had vision of flood management by embanking the rivers.

The farmers of Bengal learnt to live with the flood, and they were aware of the ecological roles of the flood, especially, the sediment dispersal and restoration of fertility in agricultural field. But the colonial rulers started to embank rivers since late eighteenth century with a view to controlling floods which often damaged agricultural production as well as delinked communication lines. Some scholars were involved in research explaining the relation between the dynamics of the river system and the human society (Willcocks 1930; Mukherjee 1938; Majumdar 1942; Bagchi 1944). The University of Calcutta came forward to publish these outstanding researches. Some other eminent scholars expressed their anxiety at the distress of the afflicted community during floods and criticized the colonial model of road–railway network that was superimposed on the inappropriate physiographic setting of the Ganga–Brahmaputra delta (Saha 1933; Ray 1932; Chapman and Rudra 2007). The colonial model of development in Bengal was guided by the logic of profit and the engineering of command and control over nature which ultimately caused an ecological rapture. The British left India in 1947 but the flawed western style of river management continues till date (Rudra 2015).

1.1 The Ganga–Brahmaputra–Meghna (GBM) Delta

The Ganga–Brahmaputra–Meghna delta is most dynamic fluvial system of the world. The tract looks like Greek letter Δ and its boundary is marked by the two major braches in the west and in the north is generally called Delta. The Meghna estuary delineates the south-eastern border, and the Bay of Bengal lies in the south. The alluvial plain covering an area of 57,000 km^2 was popularly described as Ganga delta (Rennell 1780; Bagchi 1944). A group of modern earth scientists treats the entire plains of Bengal as GBM delta (Goodbred Jr and Kuehl 2000; Goodbred Jr et al. 2014). In this newly formed landmass, fluvial and marine land building processes are working together. It seems to be an apparent paradox of the nature that in spite of influx of the substantial sediment load by the Ganga–Brahmaputra–Meghna Rivers, the land has not been growing southward along its western littoral tract rather the sea encroached inland. (Rudra 2014). But the Meghna estuary in the east recorded appreciable accretion (Sarkar et al. 2013). The geographical boundary of GBM delta almost coincides with tectonic unit known as the Bengal basin which covers 200,000 km^2. This area can be subdivided into (a) tidally active delta or Sundarban, (b) fluvial delta lying north of tidal regime (c) fan delta or Barind tract, (d) Sylhet basin and (e) lower Meghna region presently receiving combined sediment load of GBM system. The processes leading to the change in the geography of the delta have been operating in many fronts. The off-take where the Ganga is bifurcated into two major distributaries, the Bhagirathi–Hugli, and the Ganga/Padma has migrated towards the south-east from Suti to Mithipur (distance of about 16 km) in Murshidabad district. Since the late eighteenth century, all rivers have experienced rapid sedimentation and the Bay of Bengal has encroached inland along the southern coast. This shrinking of the delta can best be understood by comparing multi-dated maps and recent satellite images (Rudra 2012). The delta is reduced by about 420 km^2 along the coast during the period 1917–2016 but it continues to grow under the water. The upper or northern part of the subaqueous or underwater delta, as seen in satellite image, covers more than 25,000 km^2. The Meghna estuary is ultimate conduit of sediment influx of the GBM system into the Bay of Bengal, and consequently accretion has been a fast process leading to emergence of new land.

The processes leading to accretion and erosion of the delta are multi-faceted. The dispersal and deposition of sediment load take place during the flood when water spills off the bank. This intermittent process is active in northern non-tidal regime. In the southern littoral tract, the dispersal of sediment on the floodplain is a diurnal event linked with each high tide. This process of accretion and land building has gradually shifted southward with the growth of delta during the Holocene period (12,000 years BP). But this was largely interrupted since the late eighteenth century when the western part of Sundarban was reclaimed and tidal creeks were embanked to prevent dispersal of silt-laden saline water into the floodplain. The long-term impact of that premature reclamation is now realized. Since the process of accretion was never hampered in the eastern non-reclaimed part lying within India, the area now stands about two metres above the level of its western counterpart. The creeks of western Sundarban have been trapped within the embankments, and the sediment loads were deposited within wetted perimeter resulting in continuous rise of the bed level. The creeks which are embanked in reclaimed western part now stand above surrounding floodplain during high tide. This has created serious drainage congestion in this area. Bangladesh has built 4000-km-long embankments since 1960s and 30 million people live in the polders (the land enclosed by embankment) in Bangladesh Sundarban. The process of sediment dispersal is also interrupted in this area. The century-scale changes in both the Hugli and the Meghna estuary are striking. While the sea continuously encroaches inland along the Hugli estuary, the Meghna estuary experienced significant accretion and formation of new land.

1.2 Dynamic River System

The Ganga, the Brahmaputra and the Meghna nourish Bengal with their silt-laden water (Table 1.1), and the rivers may be divided into five major groups: (i) the rivers of Barind tract (ii) the Ganga–Bhagirathi system; (iii) the rivers of Rarh Bengal; (iv) the Jamuna (Brahmaputra)–Meghna system; and (v) the tidal creeks of littoral tract. Each group of rivers is distinctly different from the others in hydro-geomorphic characteristics (Figs. 1.2 and 1.3).

In this alluvial plain, the rivers have the tendency to move to and fro within the meander belts which may also be described as the cyclical oscillation. The rivers often encroach towards villages on the bank when a larger volume of water flows during the monsoon. In fact, many villages have developed within the meander belts of rivers and thus have been inherently vulnerable to erosion. It has not been possible to determine the period required to complete a cycle of lateral oscillation because it differs from case to case, and human intervention in the form of bank protection work retards the cycle. The other form of dynamism is the avulsion which may happen during a devastating flood, or the process may be slow covering a protracted period. In the deltaic plain, a singular channel may be bifurcated into two or more distributaries when the slope declines below a critical limit. The river adopts one of its multiple distributaries as the principal outlet of water and that may be guided by the neo-tectonic of the basin. The eastward migration of Teesta in 1787, the southward detour of Damodar in mid seventeenth century, the westward shift of the Brahmaputra or Jamuna in 1830 and the eastward flow of the water of the Ganga through the Padma (name of the main distributary of the Ganga in Bangladesh) are four examples of avulsions (Rudra, 2008). The Mahananda has recently adopted Fulohar as its principal outlet to the Ganga at Manikchak of Malda district (West Bengal), and consequently the older outlet (which joins the Ganga at Godagarighat of Bangladesh) has been reduced appreciably. The river generally leaves behind moribund channels after the avulsion. Many palaeo-channels of the delta remind us of the dynamics of the river during the historical period (Rudra 2012).

Table 1.1 Ganga–Brahmaputra–Meghna basins

River	Length in km	Catchment in km^2	Annual flow in BCM
Ganga	2510	1,087,300	525
Brahmaputra	2900	552,000	700
Meghna	210	82,000	150

Source Rivers Beyond Border, IUCN (2014)

1.2 Dynamic River System

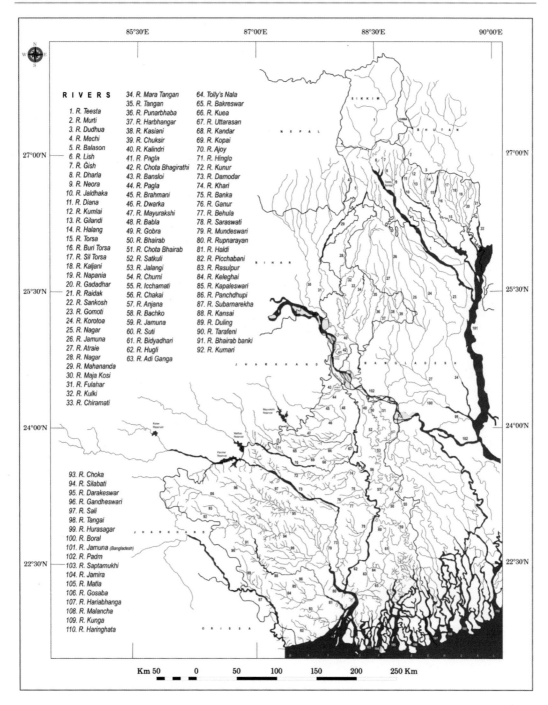

Fig. 1.2 Rivers of West Bengal

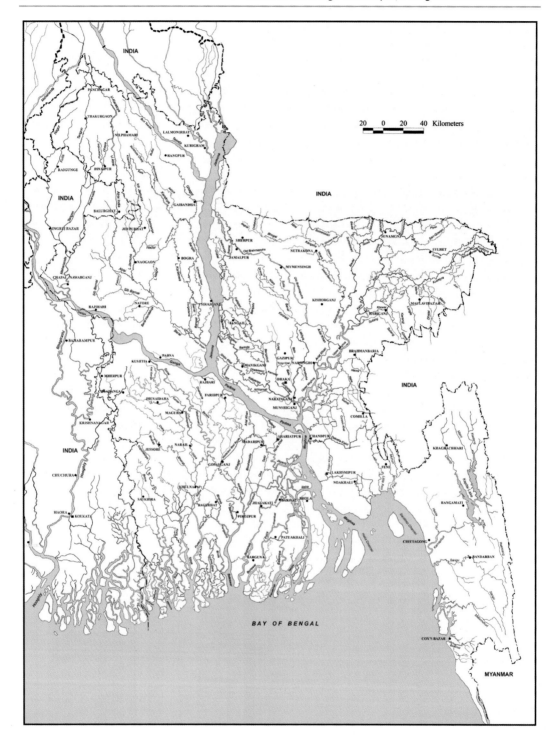

Fig. 1.3 Rivers of Bangladesh. *Source* Modified after BDPUB (2011)

1.3 Rivers of Barind Tract

The Ganga and the Brahmaputra enter Bengal through gap between the Rajmahal hill and the Meghalaya plateau. The interfluve between these two rivers is known as Barind tract and is drained by the Mahananda, the Teesta, the Jaldhaka, the Torsa and many smaller rivers. While the Mahananda discharges into the Ganga, the remaining three go to the Brahmaputra. The south-west monsoon strikes the Himalaya and generates huge rain, and this as well as snowmelt water replenishes numerous rivers draining the southern slope of the Himalaya and ultimately discharging either into the Jamuna (Brahmaputra) or into the Ganga/Padma (Fig. 1.1).

The rivers debouch on the plains from the Himalaya roughly at an elevation of 300 m, becoming wide and sluggish, and form a triangular depositional landform called fan. The plains of Tarai and Doors of North Bengal are virtually coalescing fans. The larger rivers like the Mahananda, the Teesta, the Jaldhaka and the Torsa have formed overlapping depositional lobes representing multi-dated sediment layers. The Barind tract is a fan which was formed at the end of Pleistocene ice age when the Ganga–Brahmaputra carried huge rain and snowmelt water with substantial sediment load. However, the flow in the rivers of North Bengal generally reaches the maximum level in the month of August when a total volume of more than 15,000 million cubic metres of water flows. The shallow cross-sectional areas of these rivers can hardly accommodate the huge monsoonal discharge, and floods are almost annual events in North Bengal. Where a river drops on the plain, the sudden loss of energy and resultant deposition on the bed lead to the widening and braiding of the channels. The channel form is governed by available discharge, sediment load and declining slope.

The first major human intervention altering the hydro-ecology of North Bengal was initiated from 1835 onwards when large tea estates were created at the expense of dense tropical forest. The second structural intervention that changed the hydro-dynamics of rivers was the linear embankments which were built to control floods. But this popular measure did not ensure total protection against floods; on the contrary, the sediment dispersal on the floodplain and deepening of channels by fast-flowing current were impaired. As usual, the sediment load was trapped between the embankment leading to decay of channels and resultant drainage congestion. The third intervention in the fluvial regime came in the form of narrow bridges and culverts. The road–railway networks in North Bengal is aligned in an east–west direction which cross the south-flowing rivers, and the bridges constructed for the purpose did not provide adequate passage for the floodwater. The subsequent impact of these bridges on the fluvial system was striking. In all cases, the channels became wide, both upstream and downstream, and the bridges acted as nodal points. This ultimately caused expansion of the flood contour (Ray 1932; Saha 1933).

The drainage map of North Bengal has changed since the late eighteenth century. This can be understood with reference to the maps of James Rennell who made the first systematic survey of the Ganga–Brahmaputra system during 1764–1777. A compilation of thirteen maps entitled 'A Bengal Atlas' came out of the press in 1780, and eight additional maps were added in the second edition of 1781. A look at Rennell's map and comparison of the same with the modern satellite images reveal how both the Teesta in North Bengal and the Jamuna or Brahmaputra in Bangladesh have altered their courses. The rivers of North Bengal except the Mahananda have been flowing towards the south-east, and this might be guided by a subsidence along a line presently followed by the Jamuna in Bangladesh (Morgan and McIntire 1959). While all rivers flow south-eastwards, the Mahananda exceptionally flows south-west. It debouches on the plains from the Darjeeling Himalaya near Siliguri and flows south-westward into Bihar and is bifurcated into two branches at Bagdob; the western branch is called the Fulohar which joins the Ganga near Manikchak Ghat in Maldah. The other branch known as the Mahananda itself is virtually disconnected from its feeder and does not receive any upstream flow except during high flood. The water contributed by the Balason and other Himalayan rivers now flows through the Fulohar branch. The eastern branch or the Mahananda receives the Nagor, the Kulik, the Chiramati, the Tangon and the Punarbhaba as the left bank tributaries. A branch of the Ganga named the Kalindri carries excess floodwater and joins the Mahananda at Nimasarai Ghat at Maldah. This channel had been the principal passage of the Ganga water till the fourteenth century (Mukherjee 1938). When the Kalindri carried the principal freshet of the Ganga, the lower the Mahananda below Nimasarai Ghat was a part of it (Fig. 4.7). The Mahananda basin which earlier drained an area of 27,654 km^2 is now bifurcated into two separate sub-basins which are called Fulohar and Mahananda. This change has gradually taken place during preceding three decades.

1.4 The Ganga and the Bhagirathi–Hugli

The Ganga enters West Bengal near Rajmahal hills and flows for about 56 km in the south-easterly direction to reach Farakka. The Ganga is divided into two major distributaries, the Padma and Bhagirathi, at a place called Mithipur, about 40 km south-east of the Farakka barrage. The Ganga flows another 92 km towards east before it leaves West Bengal at Jalangi which is the last major settlement along Indo-Bangladesh border. The Ganga threw off three distributaries, namely Kalindri, Chhoto Bhagirathi and Pagla, all from the left bank between Rajmahal and Farakka but further downstream, all major distributaries of the Ganga including the Bhagirathi, the Bhairab, the Jalangi, the Mathabhanga and the Garai take off from the right bank. The Boral which originates from the left bank of the Ganga at Sarda (a few km east of Rajshahi) and flows through Hura Sagar to the Jamuna had been an important spill channel of the Ganga in the late eighteenth century. The Ganga or Padma had been flowing independently through the passage presently followed by Arial Khan–Tentulia estuary.

The presently active combined outlet of the Ganga, the Jamuna and the Meghna was opened in 1830 when the Jamuna or Brahmaputra avulsed from its old course through Mymensingh and adopted a minor channel called Janai. All the distributaries of the Ganga remain beheaded during lean months and do not receive any upstream flow. The Ganga along with its tributaries and distributaries constitute the largest fluvial system draining Bengal.

The Ganga has been oscillating within wide limits and also discharged through many distributaries to facilitate water and sediment dispersal. The channels have been periodically rejuvenated and decayed. The old courses of the Ganga are now left as moribund channels. In an uncontrolled situation, the Ganga enjoyed the opportunity of free swing and the limit of swing was determined by geological outliers, i.e. older and comparatively resistant geological formation. Since the Farakka barrage was constructed and 87 million cubic metres of water was impounded, the river tried to adjust with the new hydraulic regime. The various bank protection works exerted appreciable impacts on the behaviour of the Ganga. Besides the 2.64-km-long Farakka barrage, other major structural interventions include constructions of the embankments, protection of bank with the boulders, constructions of spurs which may deflect impinging current. But most of these bank protection works proved futile. The mighty Ganga continues to impinge its bank with immense power during the monsoon and causes damage to human settlements. The changing course of Ganga in Maldah and Murshidabad has rendered thousands of people homeless (Rudra 2006, 2009). The lower reach of the Ganga in Bangladesh is known Padma which swallows fertile land and villages in every monsoon season and so described as 'destroyer of creation'.

The Bhagirathi–Hugli is the only distributary of the Ganga that has some tributaries of its own. It takes off from the Ganga and flows southward for 500 km and ultimately discharges into the Bay of Bengal at Gangasagar. The lower 280 km of this river is tide-affected. The shrine of Nabadwip marks the northern tidal limit. The two other distributaries of the Ganga, Jalangi and Churni, join the Bhagirathi at Mayapur and Shibpur Ghat near Pairadanga, respectively. The combined flow takes the name Hugli in lower reach. Kolkata emerged as an inland port during the late eighteenth century under the British rule, but the depth was very shallow and only light wooden ships could sail in the river. The steel-made ships requiring greater draught started to operate in the Hugli estuary during subsequent centuries. But the rapid sedimentation in the river impeded free movement of ships. This deepened the anxiety of port managers. The navigation in the Hugli River had been tide-dependent and difficult since inception of the port of Kolkata. The Hugli is a macro-tidal and funnel-shaped estuary. The water level may go up more than 5 m during high tide when a large ship can ply in the river. But navigation is virtually impossible during low tide. The Farakka barrage was commissioned to flush the sediment load to the deeper sea but fourteen bars or shoals continue to impede navigation in the estuary. The Hugli is a north–south aligned funnel-shaped estuary where tidal water overrides freshwater flow. It has been observed that more than 26 million ton of suspended solids remain in suspension with flowing water between Gangasagar and Diamond Harbour. Kolkata Port Trust dredges the navigation channel to ensure free passage to seagoing vessels. But the navigation continues to be an extremely difficult task (Fig. 1.2).

1.5 The Rivers of Rarh Bengal

The undulating terrain with adjoining plain lying west of the Bengal basin is known as Rarh Bengal. The tributaries of the Bhagirathi–Hugli River draining Rarh Bengal are the Bansloi, the Pagla, the Mayurakshi, the Ajay, the Damodar, the Darakeswar, the Silai, the Kansai, the Rupnarayan and the Haldi. Many of these rivers have their sources in the Chotanagpur plateau beyond the western border of West Bengal. These rivers are exclusively rainfed and flow eastward to join the Bhagirathi–Hugli River. The tributaries draining western uplands have a total catchment area covering more than 67,000 km^2. These tributaries are of different length and magnitude. The Rarh tract is divided into two topographic units, (a) the undulating terrain and (b) lower plain. These two topographic units are differentiated by 18-m-contour line. The larger magnitude of the upper unit makes the lower one flood-prone. Broadly speaking, these rivers are east flowing and tend to flow south-east in the lower reach. Some larger rivers have changed their courses in lower reaches. The Damodar which earlier discharged into Bhagirathi–Hugli River through many distributaries has changed its course and now flows through the Amta channel and the Mundeswari. The palaeo-channels can still be traced in the field and also in the satellite image as chains of stagnant pools (Rudra 2012).

1.6 The Brahmaputra-Meghna System

The Brahmaputra takes a southward bend from Dhubri (Goalpara district, Assam) and enters Bangladesh where it is known as Jamuna. The Jamuna flows through numerous braided channels down to Aricha where it combines with the Padma. The mighty river throws off a distributary called the old Brahmaputra at Dewanganj (in Jamalpur district) and that moribund channel flows through Mymensingh to join Meghna at Bhairab Bazar (in Kishoreganj district). This channel had been the main flow of the Brahmaputra before its westward avulsion in the early nineteenth century. The Barak, the main headstream of the Meghna originates from the northern hills of Manipur and flows about 560 km before reaching Karimpur district where then it is divided into two branches, the Surma and the Kusiyara; these two channels again join together at Kuliarchar (in Kishorganj district) and one named as Meghna. However, the Meghna receives the Padma at Chandpur and flows through the largest estuary flowing along Bhola Island. This is the route through which the Ganga, the Jamuna and the Meghna systems annually discharge 1350 BCM of water into the sea. The Meghna system along with the old Brahmaputra flows through the slowly subsiding Sylhet basin of north-east Bangladesh (Fig. 1.3).

1.7 The Tidal Creeks

The southern littoral tracts of Bengal, popularly known as the Sundarban, are drained by interlacing networks of numerous tidal creeks. Most of these creeks do not have any upstream supply of water, and they are only fed by tidally induced water from the Bay of Bengal. These creeks are also treated as a part of the Ganga system, and so the river is described as *Satamukhi* (one with a hundred mouths) in the Mediaeval Bengali literature. The channels flowing through the Sundarban are not rivers in strict sense of fluvial geomorphology. A river is supposed to have a source, outfall, catchment area, many tributaries and distributaries. Unlike the general nature of rivers, the creeks receive water supply from downstream, that is, from the sea or ocean. The flow is governed by the tide-velocity asymmetry and thus is a two-way flow. Most channels in Sundarban can be broadly identified as tidal creeks, with a few exceptions like Ichhamati–Haribhanga in India, the Gorai–Madhumati–Baleswar and the Arial Khan–Tentulia in Bangladesh. These three distributaries receive a feeble supply of freshwater during the monsoon months. The channel networks across the Sundarban are mostly tidal inlets and intermediate distributaries that have been beheaded or disconnected from the Ganga or Padma, having no upstream freshwater supply except the Hugli in the west and the Meghna in the east.

About 15% of the estimated one billion tonnes of sediments are sequestered in Bengal basin, and the rivers do not carry them down the estuary (Goodbred Jr and Kuhel 1998). Since a large part of the Sundarban stands less than 3 m above mean sea level and the tide rises to around 5–6 m when spring tides are coupled with cyclonic storms, the water tends to spill off the banks and recede at the time of low tide. This happen mostly in the months of August to September. The silt-laden water travels up to the northern tidal limit of the creeks. Thus, a process of accretion operates to build up the floodplain. The deposition of silt gradually blocks the flow of the river, and the old creek splits up around obstructions, thus forming numerous channels, tidal creeks and distributaries. The cross-channel connecting two larger creeks face faster decay due to head on collision of the tidal waters leading to deposition of sediment load.

1.8 Managing the Rivers

(a) Colonial Period

The flowing waters in rivers of Bengal during the late monsoon exceed the critical limit and submerge the adjoining floodplain. The people of Bengal have been living with floods from time immemorial. They learnt the art of utilizing fertile silt deposited on agricultural fields during the floods. The agricultural prosperity of Bengal was linked with the ecological services of the rivers. Since the nineteenth century, the British rulers in collaboration with landlords planned to achieve freedom from flood and built earthen embankments along the banks of rivers which had the tendency to spill over during the monsoon. The main objective of jacketing rivers was to protect the interest of landlords who used to pay annual revenue to the treasury of the British East India Company. The rivers which were initially sought to bring under control included the Kansai, the Silai, the Dwarkeswar, the Ajay, the Mayurkashi and the Damodar; the latter was designated as the 'sorrow of Bengal'. Such a colonial attribute to the Damodar was really a misnomer because the lower Damodar region ranked high in agricultural production and the upper part of the basin contained huge coal reserve. But the agricultural prosperity gradually declined since the river was embanked. The embankment or *bunds* ensured protection from low-intensity floods, but trapped the sediment load in riverbed, causing the decay of drainage systems and consequent crisis. The *bund* or embankment could not prevent high-intensity flood, as frequent raptures of the embankments and overtopping could not be avoided. The concept of security against floods provided by the *bund* ultimately proved futile. Neither the *Zamindars* (landlords) nor British Engineers could foresee the hydro-geomorphological impacts of embanking the rivers in the long run. The programme of taming the rivers gradually extended from the plains of North Bengal down to the Sundarban. The rivers in plains of North Bengal are shallow and wide. These rivers were also jacketed but floods continued to imperil human society. The ecological degeneration relating to embankment building was manifold but the most important was that the farmlands were deprived from the annual deposition fertilizing silt. This was described, in *Rarh* Bengal as the red-water famine (Mukherjee 1938). The Bardhaman district was so gifted with the silt of Damodar that it ranked first in agricultural productivity in the entire country (Hamilton 1820).

The condition deteriorated further as the European railway-road model of development was implanted on the inappropriate physical condition of Bengal and continued to expand further (Chapman and Rudra 2007). The control over Bengal by the British after the battle of Plassey in 1757 was contemporaneous with the Industrial Revolution in Britain. The availability of cheap coal, iron/steel and the invention of steam power set the stage for the railway revolution from the 1830s. The investment in railways was huge but thought to be secure. There was a long discussion in Britain over the relative efficacy of railway or canal. The engineers who advocated the improvement of navigation in Britain were blamed as suffering from *canal mania*. Considering the intricate network of rivers, a group of engineers, especially Sir Arthur Cotton, advocated the development of navigation in Bengal which was then one of the main sources of raw materials for industrial growth in Britain. The rail and road lobby was ultimately successful in convincing the policy-makers because of their stronghold in the British Parliament. The Sepoy Mutiny or the first war of independence in 1857 generated an idea that roads and railways would be the best means of communication in India to deploy the army quickly to the places of revolt. The dense tropical forest of the Chotanagpur plateau was indiscriminately cleared to provide support (sleepers) under railway tracks. This caused more erosion of topsoil in deforested areas and increasing sediment load in rivers. The roads and railways were built on high embankments to ensure uninterrupted movement even during the flood. The earth materials for building embankments were borrowed from ditches parallel to the lines of communication, and those borrow pits subsequently became linear swamps and breeding grounds of mosquito. The spread of malaria followed the paths of railway and highway. The construction of the railway from Sealdah to Lalgola created drainage congestion in the area between eastern bank of the Bhagirathi and the railway embankment causing a widespread epidemic of malaria in Murshidabad and Nadia (Samanta 2002). Since the railways and highways in both north and south Bengal were transverse to the flow direction of rivers and were built with narrow culverts, there were drainage congestions, expansion of flood contours and decline in agricultural productivity. So the embankments, roads and railways were described as *satanic chains* (Willcocks 1930). In fact, the cost of urban-industrial living was externalized on rural Bengal. Saha (1933) explained the ecological degeneration of the lower Damodar plain after the East India railway connected upper India with Kolkata. Two years after opening of the railway in 1859, malarial epidemics caused the death of about one million people in the Hugli district and the fertility of soil declined in both Bardhaman and Hugli. The case of North Bengal was not an exception. The devastating flood of 1922 was largely due to the east–west aligned railway which intercepted south-flowing rivers (Ray 1932).

The colonial hydrology of India has been discussed by many authors. While explaining the relationship between colonialism and water, De'Souza (2006) identified three overlapping but discrete areas of concerns. These are colonial irrigation strategies, decline, elimination and appropriation of traditional irrigation technologies and hydraulic endowment. The last area of concern encompassed the colonial attitude to river management in respect of flood, inland navigation and holistic river management. The British Engineers had the intention of developing a network of canals so that farmlands could be irrigated even during lean months. For this, it was necessary to store monsoon water and transfer the same to the non-monsoon season. A long canal connecting Kansai in Medinipur to Damodar at Uluberia was excavated with the twin purposes of irrigation as well as navigation. It was opened for traffic in 1863. A weir on Kansai at Medinipur was built in 1869 to induce water in the canal. The canal was designed without any prior estimate of the discharge that flowed through the river and the area that could be actually irrigated. So the irrigated area fell far short of the target. The canal with high banks created drainage congestion (Inglis 1909). The Orissa Coastal Canal which was excavated with the vision of connecting Kolkata and Puri during 1880–1886 was protected with high embankment on both sides and that filled up many channel draining to the sea (Ray 1932). The water harvesting and irrigation in Bengal, during pre-British days, were indigenous. It was a custom to excavate large tanks in drought prone areas of *Rarh* Bengal. But by early twentieth century, those tanks were silted up and either became rice fields or stagnant pools of polluted water causing the spread of cholera or malaria (ibid).

The East India Company started premature reclamation of the western part of Sundarban in the late eighteenth century when land building in the coastal region was incomplete. The clearing of forests along with creating polders in Sundarban imperilled the delicate hydro-ecological balance, but the flawed logic of protecting the area with embankment still persists in engineering mindset. The embankment did not ensure protection against storm surge. Cyclones or tidal surges frequently caused the increased volume of water to spill over the embankment and that water did not find any way to flow back to the river. However, the sediment dispersal and delta building continued uninterruptedly in the non-reclaimed parts of Sundarban where sediment-laden tidal water got the free opportunity to spill off and being deposited on the floodplain. This has led to the rise in elevation of the non-reclaimed forested lands as compared to the reclaimed areas. More than 12 million people living in undivided Sundarban continuously struggle for existence. It is necessary to provide more space for the rivers and rebuild the embankments away from high tide limit. But intertidal spaces are so densely occupied that it is extremely difficult to evacuate so many people.

(b) Post-colonial Period

The spatially and seasonally uneven rainfall has compelled the water managers, since the dawn of civilization, to store water during the monsoon and transferring it to the non-monsoon season when society suffers from acute shortage of rainfall and minimum flow in rivers. The engineers, entrusted to manage our river system in independent India, borrowed the philosophy of 'command and control over the hydraulic system' initially from British Institutions and subsequently from the USA. The Damodar Valley Corporation was designed keeping the Tennessee Valley Authority as the model. The British Engineers had an incomplete understanding of the hydro-geomorphology of the Bengal Delta where farmers had invented the methods to make the best possible utilization of the natural system. In particular, the role of silt-laden floodwater in agriculture was not understood by the colonial rulers.

The structural interventions across the rivers, the excavation of canals to withdraw water, and the building of linear flood control embankments, created discrete control systems which impaired the delicately balanced and integrated river system of Bengal. The structures built across Damodar at Rhondia or Kansai at Medinipur were low and allowed the flowing water an opportunity of overtopping the barriers. But the post-independence era witnessed massive structures which often left the downstream stretch of the river totally dry. It was understood too late that many ecological services had ceased subsequent to the taming of the rivers. The adopted management policies invited many problems. Since the engineers had no understanding about the transmission–distribution loss of water in dam–canal network, the gap between targeted irrigation and the area truly irrigated gradually widened and there was increasing dependence on ground water to fill up this gap. The exploitation of groundwater exceeding replenishable limit led to the diminution of base flow and ultimately desiccation of rivers. Neither the flood control nor the targeted power generation was achieved by the Damodar Valley Corporation. The experiences of two other projects of South Bengal, namely the Mayurakshi and the Kanshabati and the Teesta barrage Project in Jalpaiguri (West Bengal) and another at Dalia (Bangladesh) were similar to that of the DVC. It was found impossible to generate hydroelectric power and ensure water for irrigation from a single reservoir as power generation requires uninterrupted supply of water to rotate turbine and irrigation needs to transfer water from one season to other. However, the demand for irrigation sector was so high that no water was virtually left in rivers for ecological services.

1.9 Ecological Flow

The importance of water in the environment is now accepted by all (Gleick 1993). It is important to understand the threshold limit of altering fluvial regime. The arithmetic hydrology determines ecological flow in rivers in terms percentage of total flow. The concept of ecological flow identifies the required quantity, quality and distribution of water from the source to the mouth in a river, preserving the life in and around it. It also recognizes the importance of environmental flows 'as a vital contributor to the continuing provision of environmental goods and services upon which peoples' lives and livelihood depend' (WWF 2012). The term ecological flow also includes its cultural and aesthetic value. It identifies the required quantity, quality and distribution of flow patterns from the source to mouth of a river, preserving the life in and around the channels (Bandyopadhyay 2011). The drastic reduction in flow in most of the river has the ecological cost. An already identifiable indicator of the ecological damage is the falling level of the groundwater table in the downstream of dams/barrages and the loss of biodiversity. The greater common good of the ecosystem and human society does not lie in the abuse of the rivers; the future of civilization depends on fixing critical limits for the exploitation of water sources and allowing the rivers to flow.

References

Bagchi K (1944) The Ganges Delta. University of Calcutta, Calcutta

Bagge A (1950) Report of the international arbitral awards. Boundary disputes between India and Pakistan relating to the interpretation of report of Bengal Boundary Commission. Part I Published by UN (available on line)

Bangladesh Pani (Water) Unnayan (Development) Board (2011) Rivers of Bangladesh (in Bengali)

Bandyopadhyay J (2011) Deciphering environmental flows. Seminar 626:50–53

Chapman GP, Rudra K (2007) Water as Foe, Water as Friend: lesson from Bengal's Millenium Flood. J South Asian Develop 2(1):19–49

De'Souza R (2006) Drowned and Dammed/Colonial Capital and Flood Control in Eastern India. Oxford, New Delhi

Gleick Peter H (1993) Water in crisis: a guide to world's fresh water resources. OUP, Oxford

Goodbred SL Jr, Kuehl SA (1998) 'Floodplain processes in the Bengal Basin and the storage of Ganges-Brahmaputra river sediment: an accretion study using 137Cs and 210Pb geochronology. Sed Geol 121 (1998):239–258

Goodbred SL, Kuehl SA (2000) The significance of large sediment supply, active tectonism, and eustasy on margin sequence development: Late Quaternary stratigraphy and evolution of the Ganges-Brahmaputra delta. Sed Geol 133:227–248

Goodbred Jr SL, Penny MP, Ullah MS, Pate RD, Khan SR, Kuehl SA, Singh SK, Rahaman W (2014) Piecing together the Ganges-Brahmaputra-Meghna River delta: use of sediment provenance to reconstruct the history and interaction of multiple fluvial systems during Holocene delta evolution. Bulletin, Geological Society of America

Hanilton W (1820) Geographical, Statistical and Historical Description of Hindoostan and the Adjacent Countries, vol I. John Murray, p 157

Inglis WA (1909) The Canals and Flood Banks of Bengal. Bengal Secretariat Press, Reprinted in Rivers of Bengal, V(I):62–83, West Bengal district Gazetteers (2002)

IUCN (2014) Rivers Beyond Border. India Bangladesh Transboundary River Atlas
Majumdar SC (1942) Rivers of the Bengal Delta. University of Calcutta
Morgan JP, McIntire WG (1959) Quaternary Geology of the Bengal Basin, East Pakistan and India. Geological Society of America Bulletin 70(3):319
Mukherjee RK (1938) The Changing Face of Bengal. University of Calcutta, Reprinted (2009)
Ray PC (1932) Life and Experience of a Bengali Chemist, II. pp 159–160, Reprinted by Asiatic Society(1996), Kolkata
Rennell J (1780) A Bengal Atlas containing maps of the theatre of war and commerce on that side of Hindoostan. London. (edited by Kalyan Rudra(2016) and reprinted by Sahitya Samasad, Kolkata)
Rudra K (2006) Shifting of the Ganga and Land Erosion in West Bengal/A Socio-ecological viewpoint. CDEP Occasional Paper 8. IIM Kolkata
Rudra K (2008) Banglar Nadikatha (in Bengali). Sahitya Samsad, Kolkata
Rudra K (2009) Dynamics of the Ganga in West Bengal, India (1764–2007). Implications for science. Quatern Int 227(2):161–169. https://doi.org/10.1016/j.quaint.2009.10.043
Rudra K (2012) Atlas of Changing River Courses in West Bengal. Sea Explorers' Institute, Kolkata
Rudra K (2014) Changing River Courses in the Western Part of the Ganga-Brahmaputra Delta. Geomorpholgy 227:87–100
Rudra K (2015) Rivers of West Bengal/Dying, Living. In: Ramaswamy R (ed) Living Rivers/Dying Rivers. Iyer. Oxford, New Delhi, pp 188–204
Saha MN (1933) 'Need for a Hydraulic Research Laboratory', Reprinted (1987) in Collected Works of Meghnad Saha. Orient Longman, Calcutta
Samanta A (2002) Malarial Fever in Colonial Bengal/1820-1939/Social History of an Epidemic. Firma KLM Pvt. Ltd., Kolkata
Sarkar MH, Akter J, Rahman MM (2013) Century-Scale Dynamics of the Bengal Delta and Future Development. In: Fourth international conference on water and flood management, pp 91–104
Willcocks W (1930) Ancient System of Irrigation in Bengal. University of Calcutta
WWF India (2012) Report on Assessment of Environmental Flows for the Upper Ganga Basin. (downloaded from http://awsassets.wwfindia.org/downloads/wwf_e_flows_report.pdf)

Evolution of the Bengal Basin

Abstract

This chapter deals with the geological evolution of the Bengal basin in the context of plate tectonics in the South Asian region. The basin contains thick early Cretaceous–Holocene sedimentary successions. It is estimated that half a million kilometre^3 of sediment has been deposited so far, leading to formation of the largest delta and associated plain land of the world. The basin filling and delta building got an accelerated momentum during late Holocene period when the Himalayan glaciers decelerated contributing huge sediment loads into the GBM system. The Bengal basin has experienced recurrent marine transgression and regression during preceding geological ages. The fluvio-marine process of sediment accretion as we see today in Sundarban has gradually shifted southwards during last 20 million years. The coastline along the Indian Sundarban in the west has been encroaching inland due to subsidence of land and rising sea level but the Meghna estuary in Bangladesh has been growing southwards due to accretion.

The vast alluvial plain with Barind and Madhupur terrace as higher lateritic tracts and interlacing channels of the Ganga–Brahmaputra and the Meghna system have given the Bengal basin a special geomorphic identity (Fig. 2.1). It is the cradle of the Ganga–Brahmaputra–Meghna (GBM) delta and characterized by many palaeo-channels and shallow swamps. The extensive alluvial plains and associated uplands are shared by India (West Bengal) and Bangladesh. It is surrounded by the Indian craton in the west, the Meghalaya plateau and the frontal plains along Himalaya in the north and the fold mountains of Manipur–Tripura–Chittagong in the east and covers an area of 200,000 km^2. A group of experts belonging to an older school delineated the triangular tract encompassed by the Bhagirathi-Hugli in the west, the Padma in the north, combined flow of the Padma, the Jamuna and the Meghna in the east and Bay of Bengal in the south as the Ganga delta (Rennell 1780; Bagchi 1944). Considering the role of the Brahmaputra in delta formation, the scholars of modern school treat entire alluvial plain of Bengal as the Ganga–Brahmaputra (GB) delta (Allison et al. 2003; Bandyopadhyay 2007). But it may be more correct to call it Ganga–Brahmaputra–Meghna (GBM) delta as the latter also played an appreciable role in the depositional process. In fact, three separate deltas were coalesced to form the largest delta which we call the GBM delta.

The older lateritic tracts of Barind, Madhupur and the tectonically active Sylhet basin are some striking geomorphic units composing the Bengal basin. The growth of the delta is not limited at the coastline; rather it continues under the sea

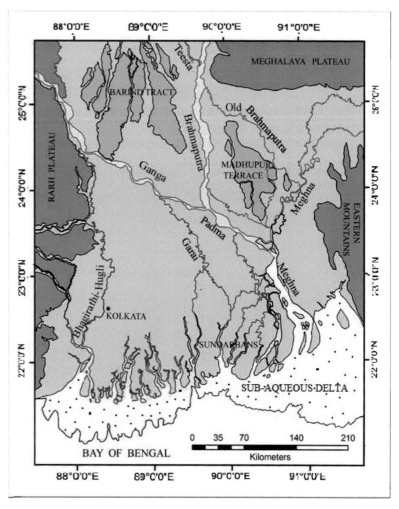

Fig. 2.1 Physiography of Bengal basin. *Source* Rudra 2012

and can be seen in the recent satellite images. The submarine delta covers 25,000 km² beyond the coast of Bengal, and the sea-floor cone of silt has reached south of Sri Lanka. The stratigraphy of GBM delta is composed with huge sand layers at depth and shallow surface silt-clay deposits because of its proximity to the southern face of the Himalaya. The rivers carrying sediment load have a comparatively short run from their sources to the sea. In contrast, the finer silt dominates Indus delta formed in the Arabian Sea and located more than three thousand kilometres south of the Himalaya (Giosan et al. 2006).

Reconstruction of the evolutionary sequences of the Bengal basin needs in-depth understanding of the plate tectonics in the context of the South Asian region. The exploration of petroleum in Bengal basin since the 1950s revealed considerable subsurface data, which helped the scientists to comprehend the basin-fill history (Sengupta 1966). The geologists also relied on seismic evidences as borehole data were inadequate to build up the complete story. The origin of the Bengal basin dates back to the early Cretaceous times (145 million years BP), when the Gondwanaland was broken up into pieces. A part of Gondwanaland which now constitutes India initially drifted north-westwards and subsequently northwards. Prior to drifting of the Indian plate, after detachment from Australia and Antarctica in early Cretaceous, the Siam, Burma, Malaysia and Sumatra (SIBUMASU) block moved northwards and confronted with the Asian block. Thus, three blocks came close to each other.

The Bengal basin was formed by the subduction of the Asian plate beneath the Burma plate. There was a soft collision (59–44 million years BP) between the Indian and Asian plates. There was also a hard collision in the early Eocene era (44 million years BP), which initiated the Himalayan orogeny. The Bengal basin became the remnant of a large ocean basin (Alam et al. 2003). Thus, an asymmetrical pericratonic rift basin was formed in the north-eastern part of the Indian plate (Mukherjee and Hazra 1997).

2.1 Structure of the Basin

The western Achaean shield which stands along western border of the Bengal basin plunges approximately at 87 °E meridian and extends further east under thick layers of recent sediment. (Sengupta 1972). The basement, on which sediment layers have been deposited, is tilted towards east. The central part of the basin in Bangladesh records nearly 22,000 m thick early Cretaceous-Holocene sedimentary successions (Alam et al. 2003). Since the basement configuration is concealed under thick sediment layers, it is difficult to distinctly identify plate boundaries and *suture zones* (areas where two plates have joined). The hinge zone, a striking structural feature of the Bengal basin, can be marked by a line connecting Kolkata and Mymensingh (Bangladesh). This structural unit, having a width of about 25 km., divides continental and oceanic parts of the Indian plate. The hinge further differentiates the stable shelf province lying to the north-west from the central deep basin province. The dip of the basement changes (from 2–3° to 6–12°) along the hinge. The dip of the rock again becomes gentler in the deeper part of the basin in Bangladesh (Uddin and Lundberg 2004). A striking feature in the western shield is the Rajmahal hills, a thick outcrop of horizontal to subhorizontal basaltic lava flow of late Jurassic or Early Cretaceous time, covering more than 4000 km^2 (Sengupta 1966). The hills and plateaus stand 150–240 metres above mean sea level though a few hills exceed 450 m in elevation. The origin of Rajmahal trap is also attributed to fracture in the crust associated with split-up of the Gondwanaland. A series of *en-echelon* faults along the western edge of the Bengal basin are probably related to deep-seated shearing movement in the basement. The Meghalaya plateau, covering an area of about 23,000 km^2, stands at the elevation between 1350 and 1800 metres, and it is identified as a pop-up structure between the Oldham fault in the north and the Dauki fault in the south (Alam et al. 2003). There are a series of compressional faults in Indo-Burma mountain ranges along the eastern margin of the Bengal basin. The origin of these reverse faults can be attributed to plunging of the Indian plate beneath the SIBUMASU plate resulting in severe compression. The folded rocks, reaching the limit of endurance, slipped westwards along the fault plane and brought older rocks over the newer ones (Figs. 2.2 and 2.3).

2.2 Basin-Fill History

The first phase of sediment filling was initiated by the western tributaries to the Bhagirathi-Hugli River. The rivers like the Mayurakshi, the Ajay, the Damodar and the Rupnarayan draining out of the western uplands formed palaeo-/subdeltas at their respective outfalls along the oldest strandline, which ran north-eastwards from Digha to Nabadwip (Agarwal and Mitra 1991). The geologists have identified four palaeo-strandlines from photo geomorphological studies, though the dates were not properly ascertained. The Damodar delta is the most pronounced and can be distinctly identified from recent satellite image. The land building in the Himalayan front (i.e. in Doors and Tarai region) of the Bengal basin started 20 million years BP in Miocene period (Banejee and Sen 1987).

It is estimated that half a million kilometre^3 of sediment has been deposited so far, leading to formation of the largest delta and associated plain land of the world. The deposition of sediment

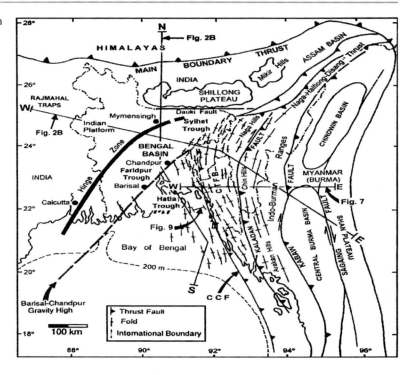

Fig. 2.2 Geological section lines across the Bengal. *Source* Alam et al. 2003

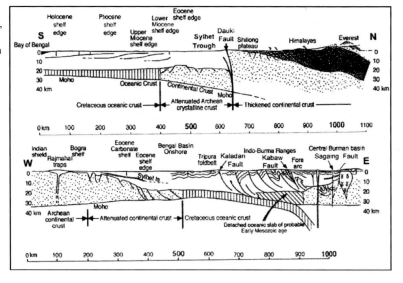

Fig. 2.3 Schematic sections across the lines N-S and E-W, showing major tectonic and crustal features. *Source* Alam et al. 2003

continues further south of the delta front or coastline. The subaqueous fan has been formed by the accretion of layers of huge sediment load, and the volume of sediment accumulated therein is about 12.5 million kilometre^3 (Curray et al. 2003). Curry (1994) divided the underwater deposits of Bay of Bengal into two parts: (a) Eocene–Holocene sedimentary formations which were deposited since the collision between India and Asia, (b) early Cretaceous–Palaeocene deposits classified as pre-collision sedimentary and meta-sedimentary rocks. The GBM delta which has emerged above mean sea level is a very small part of larger depositional landform. The *Swatch of No Ground*, a submarine canyon, is the presumed location of the river mouth during the last glacial sea level low. The combined channels of the Ganga, the Brahmaputra and the Meghna now flow to the Bay of Bengal about 90 km further east and supply little sediment into the canyon.

The greater seaward growth of the coastal tract along the western (Hugli) estuary has led to a common notion among the earth scientists that the principal flow of the Ganga passed through the western margin of the delta for a long period during the past. Though the main conduit of the Ganga-Brahmaputra-Meghna sediment lies along the eastern margin, the growth of eastern part of delta has failed to keep pace with its western counterpart. The eastward tilt of the basement seems to be the cardinal factor affecting the differential growth of delta (Rudra 1999, 2008). The greater depth of sediment layers towards the east confirms this view. The absorption of the sediment brought by the Brahmaputra in the sinking Sylhet basin till 1830 had also delayed the growth of delta along the Ganga-Brahmaputra-Meghna estuary.

2.3 Late Quaternary Sequences

Deltas are generally formed by the world's great rivers. The Amazon, the Ganga, the Hwang Ho, the Mississippi and the Nile have built massive deltas. They drain large basin areas and carry huge sediment load. Consequently, the deltas of these rivers are vast in size. The delta progradation started worldwide since the Holocene era. The investigative research in the Mississippi delta has revealed that the depocentre had been forming between 10,000 and 5000 year BP and the delta was built up by silt deposited in the Gulf of Mexico. The mouth of the Mississippi was extended further into the sea 10,000 years BP. With the onset of the Holocene period, temperature increased and consequent melting of glaciers caused the sea level to rise 5000–6000 years BP. Most of the world's deltas began to grow since the sea level stabilized. The investigation in major deltas along the Mediterranean Sea brought to light that the depocentres were activated in the period commencing from 8000 to 6000 years BP. Sixty-four radio carbon samples from basal Holocene sections of the Nile delta confirmed that the deposition took place between 8500 to 5500 years BP. It was revealed that formation of 72% of the world deltas initiated between 8000 to 6500 year BP, and the rest, from 7500 to 7000 years BP (Stanley and Warne 1994). The sea-level change and melting of glaciers acted as the cardinal factors inducing early Holocene delta development. Compilation of global sea-level data and radiocarbon dating depict notable changes of coastline along with the sea-level rising during 10,000 to 6000 years BP.

Goodbred and Kuehl (2000a) identified two major stratigraphic facies in the late quaternary sequence of South Bengal—*oxidized* and *sand*. They noted, '*oxidized facies consists of stiff silty clays brown to orange in colour and is distributed locally throughout the region at a depth of 10-45 metres. Underlying this mud up to 10 metres of weathered, iron-stained, heavy-mineral-deficient sands are also part of the oxidized facies*'. This was deposited 14,000 years BP. The basal sediments are also extensively represented by grey, micaceous, heavy-mineral-rich sand deposited prior to 8500 years BP. In the protracted period of evolution of the Bengal basin, recurring submergence and emergence had been important phases of basin-fill history. The remote part of the shelf zone presently lying within Assam, West Bengal and Bangladesh was invaded by the sea during the late Cretaceous–middle Eocene times (85–45 million years BP). The stable shelf zone or the area lying to the north-west of the hinge zone was exposed due to retreat of sea during Oligocene to early Miocene (34–20 million years BP). Uddin and Lundberg (2003) identified the stratigraphic framework of the Bengal basin in three provinces: (a) north-west Bengal basin, (b) north-east Bengal basin and (c) south-east Bengal basin.

The uplift of the mobile belt and intense deformation caused widespread recession of the sea during the late Miocene–Pliocene times. In the Quaternary period, the Bengal basin recorded further recession of the sea due to influx of huge sediment load from the GBM system.

The first phase of delta formation started 2000–3000 years earlier than the other Holocene deltas of the world. The delta took the present shape in 3000 years BP (Goodbred and Kuehl 2000a). The post-Pleistocene rise of the sea-level facilitated the formation of the Holocene deltas of the world. Stanley and Warne (1994) opined that the GBM delta started to grow since 7060 ± 120 years BP. Goodbred and Kuehl (2000a) fixed the time 10,000–11,000 years BP.

The sequence of events was as follows:

1. The sea level receded by about ∼120 m during the Pleistocene low stand, when discharge in all rivers diminished. The land was exposed to alternate wet and dry condition and thus extensive palaeosols formed.
2. Since 12,000 years BP, the gradual rise of temperature caused melting of glaciers in the Himalaya resulting in significant increase of discharge and sediment load in rivers. The Barind tract appears to be a post-Pleistocene depositional feature and overlaid by the Teesta fan (Fig. 2.4).
3. By 11,500 years BP, the south-west monsoon established strongly over this part of the globe, contributing increasing sediment load in rivers weighing 2.5 billion tonnes per year, compared to the present rate of about one billion tonnes per year.
4. By 10,000–11,000 years BP, fine-grained mud was deposited widely over oxidized and alluvial sand. The sea level went up to 75 m from the level of Pleistocene low stand causing wide spread sediment trapping. The situation continued during subsequent 2000–4000 years when sea level swelled 30–35 m at the rate of 1 cm/year. This rapid rise of sea level was counterbalanced by the huge influx of sediment caused by increasing intensity of monsoon rain.
5. The littoral (coastal) tract grew southwards during 5000–2500 years BP. In the first phase, the depocentre located in the Sylhet basin was filled up by the deposits of the old channel of Jamuna or Brahmaputra River, and the other depocentre at the Gangasagar estuary was filled up by the Hugli River and its tributaries. Subsequently a number of depocentres started operating towards the east, giving rise to the GBM delta.

The traditional hypothesis put forward by many scholars (Oldham 1870; Majumder 1942; Bagchi 1944) described the delta formation as being guided by the fluvial process. But delta building is also an outcome of the fluvio-marine processes at the river mouths where tides play the most dominant role. The tidal fluctuations and reversal flow in the littoral creeks impede the sediments influx into the Bay of Bengal and facilitate the sediment deposition at the estuarine mouth. During high tides, the silt-laden tidal water spills over the banks of the creeks and leaves behind a thick layer of alluvium on the floodplain. Thus, the fluvio-marine delta building process has slowly migrated southwards from the beginning of the Holocene period. The Bengal basin has experienced recurrent marine transgression and regression during preceding geological ages. The fluvio-marine process of sediment accretion as we see today in Sundarban has gradually shifted southwards during last 20 million years.

2.4 The Bengal Basin and the GBM Delta

The origin and evolution of the Bengal basin and the Ganga–Brahmaputra–Meghna delta is generally treated as the topic of research in geology, sedimentology or geomorphology; but truly it is a matter of interdisciplinary research including social sciences (Mukherjee 1938; Eaton 1996; Chakrabarti 2001). The research on the sedimentary history of the Bengal basin is premature till date. The Bengal basin, a geologically recent and tectonically active region, facilitated the formation of the GBM delta. Earlier the scholars described it as the Bengal delta or Ganges delta (Rennell 1780; Fergusson 1863; Majumder 1941; Bagchi 1944; Umitsu 1993; Sarkar et al. 2013). The role of the Brahmaputra and the Meghna in delta building has been recognized lately. The identification of the delta within the Bengal basin

2.4 The Bengal Basin and the GBM Delta

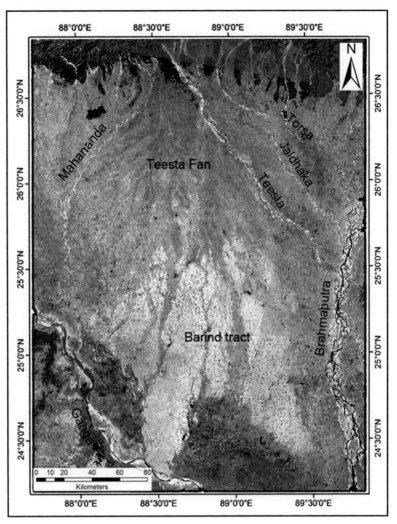

Fig. 2.4 Barind tract, a large fan

and the evolutionary processes have been debated long (Rudra 2015; Chakraborty 1970). In view of its geometrical identity, the alluvial plain lying to the east of Bhagirathi-Hugli River and south of the Ganga/Padma river is popularly called the delta (Oldham 1870; Bagchi 1944; Basu and Chakraborty 1972; Bagchi and Mukherjee 1979; Rudra 2014). The triangular area thus identified looks like Greek letter Δ. Since all the three boundaries are not fixed but the rivers change their courses continuously, the size of the delta has changed with time. The delta is expected to grow southwards due to estuarine deposition but such had not been the case along the western part where the sea has encroached inland (Rudra 2012). On the contrary, the Meghna estuary recorded fast accretion (Sarkar et al. 2013). Bandyopadhyay et al. (2015) proposed a wider extension of the GBM delta and included 'para-deltas' formed by the western tributaries to the Bhagirathi-Hugli within its geographical area. But a counterview states that the added area should not be treated as the part of the delta proper (Chatterjee 1973). The undulating plateau along with the adjoining plains lying in between the Chotanagpur plateau and the Bhagirathi-Hugli River is described as Rarh Bengal. This easterly sloping region comprises three distinct topographic units, namely the plateau fringe, the Piedmont zone and the moribund zone. The course of the Bhagirathi-Hugli flows along the proposed boundary between Rarh in the west and delta in the east (Bagchi and Mukherjee 1979).

Many recent studies applied the term delta in a wider sense and included the entire plains of Bengal as the delta (Goodbred and Kuehl 1998, 2000b; Alam et al. 2003; Allison et al. 2003; Uddin and Lundberg 2004; Sarkar et al. 2013). Bandyopadhyay (2007) divided the GBM delta into two parts, and the present course of the Ganga/Padma and the lower Meghna was treated as the dividing line. The south-western part is believed to be built by the sediment load carried by the Ganga system and north-eastern part is by the Brahmaputra and Meghna system. But the area lying to the north of the Ganga/Padma and west of Jamuna had been built by the sediment post-Pleistocene sediment load brought by the Mahananda and the Teesta and both the rivers were tributaries to the Ganga (Rudra 2015). The Teesta (which now discharges into the Jamuna) migrated eastwards during the devastating flood of 1787 (Hirst 1915; Majumder 1941; Rudra 2012). The Pleistocene deposit forming the Barind tract is delineated by the Mahananda in the west and the Teesta in the east and a post-Pleistocene depositional tract known as the Teesta fan covering an area of 18,000 km^2 overlies the Barind tract in its northern fringe (Chakraborty et al. 2010 also see Fig. 2.4). So such an concept that the Ganga/Padma divides the GBM delta diagonally and area lying north of the dividing line was built only by the Brahmaputra–Meghna system is entirely wrong.

The role of the Brahmaputra and the Meghna in the depositional processes leading to filling up of the Bengal basin had been equally important. Before the Brahmaputra finally avulsed to its present course through Bangladesh in 1830, much of sediment had been absorbed in the Sylhet basin. In the process of gradual easterly flight of the silt-laden water of the Ganga, the intermediate distributaries became active successively (Allison et al. 2003) and the Brahmaputra oscillated many times along both sides of the Madhupur tract during 18000–3000 cal years BP (Goodbred and Kuehl 2000a). The latest avulsion by the Brahmaputra was discussed in detail by Morgan and McIntire (1959) and Coleman (1969). When James Rennell conducted first systematic survey of Bengal during 1764–1777, the Ganga and the Brahmaputra had been discharging separately into the Bay of Bengal (Rennell 1780) The Brahmaputra flowed through Sylhet basin and discharged into the Bay of Bengal independently through the Meghna estuary (Hirst 1915). The recent westward migration of the Brahmaputra was not sudden but a slow process covering several decades and finally completed in 1830. A minor channel locally called Janai was adopted, and the old channel through Sylhet basin was abandoned (Fergusson 1863). Since then 'Janai' was gradually corrupted to 'Jamuna' in local dialect of Bangladesh.

Thus it can be concluded that the Ganga, the Brahmaputra and the Meghna have contributed in delta building and they had spatially different areas of working. The area enclosed by the Bhagirathi-Hugli in the west, the Ganga/Padma and the Meghna in the north and the east, and the Bay of Bengal in the south is truly the part of the GBM delta where the Ganga and its distributaries have dominated in the depositional processes. The Barind and surrounding plain had been built by the Mahananda, the Teesta and other rivers debouching from the southern slope of the eastern Himalaya. The Brahmaputra and the Meghna have filled up the Sylhet basin and associated plains. Allison et al. (2003) have identified different pathways of sediment influx into the sea and growth of lobes along the coast since 7500 years BP. While the Brahmaputra sediments were absorbed in the Sylhet basin between 7500 and 6000 years BP, the Ganga had been engaged in building three successive estuarine lobes since 5000 years BP. The process progressively moved eastward from the Hugli to the Garai estuary. The combined load of the Brahmaputra and the Meghna formed another lobe downstream of Chandpur since 6000 years BP and continued till 200 years BP. The present process is operational along Bhola Island. There are 13 major estuaries which connect the GBM system with the Bay of Bengal. The estuarine sediments are carried inland though these conduits and submergence of floodplain and sediment dispersal during high tide facilitate land building. This process had always been active in coast and shifted southwards with the growth of the delta.

The south-western part of the GBM delta is the exclusive domain of the Ganga and its distributaries and Bagchi (1944) described the area as the Ganga delta. He divided the area into three geomorphic units—moribund, mature and active delta. It is difficult to differentiate between moribund and mature delta. The distributaries in the moribund delta remain delinked from the Ganga/Padma except three monsoon months. The deltaic plains are characterized by many inter-distributary swamps and abandoned meander loops indicating past dynamics of rivers. The mature delta is juxtaposed to the moribund delta where the rivers are in alarming state of decay. The littoral tract lying further south was identified as the active delta where sediment deposition continues till day due to tidal action. In both the moribund and the mature delta, the rivers are decaying fast due to diminishing upstream flow but are rejuvenated during the monsoon rains. In the years of devastating flood layers of fresh sediment were deposited in the areas where land building was supposed to be ceased. Notably the northern part of the delta gets deposition only in the years of floods but the littoral tract is inundated twice in 24 hours due to tidal fluctuation. Since most of the creeks of the littoral tract are disconnected from the Ganga/Padma and do not receive any freshwater, this part is called tidally active delta. The sediment dispersal in the littoral tract is governed by fluvial-marine process A part of the suspended sediment load flushed into the Bay of Bengal through the estuaries is pushed back with tidal invasion and spills over the inter-tidal space.

2.5 Stratigraphy of Delta

The word delta was first used by Greek historian Herodotus, to identify the triangular depositional features which looked like the Greek letter Δ and was emerged at the estuary of the Nile River. The ideal conditions for the development of a delta are shallow continental shelf favouring deposition, a large amount of sediment influx through the river, seasonal variation of the discharge and tidal fluctuation causing regular submergence and emergence of the coastal area. It generally develops at the outfall of the river in the sea. It is generally identified as the portion of riverine deposits lying in between two marginal distributaries of the trunk stream and the sea. In such case, the apex of the delta stands at the point, where the main river is divided into two distributaries. Both the apex of the delta and the front facing the sea are dynamic and change their locations with time (Basu and Chakraborty 1972). All the east-flowing rivers of peninsular India have formed deltas at their respective outfalls. The GBM delta is unique considering its size and geomorphic characteristics, and it is virtually a coalescing delta built by three mighty rivers.

A delta may be formed when a river carrying sediments reaches a vast body of water. The hydraulic gradient is immediately reduced to near zero, hence there is a sudden reduction in velocity of flow. Additionally, when the water is no longer confined in a channel, it can expand in width. As a result it drops suspended sediments on the sea bed. The sedimentation takes place at the estuaries of the river and on the sea front. The extensive delta is formed there and gradually extends towards the sea. Generally three sets of beds are observed in the deltaic stratigraphy. These are: (a) topset beds, (b) foreset beds, (c) bottomset beds. Topset bed represents upper most bed of sedimentary layers of a delta. These are generally composed of horizontal beds or layers having gentle slope. The series of the dipping beds inclined towards the sea are called foreset beds. A horizontal layer of clay and silt deposited beyond the edge of foreset are called bottomset beds. The thickness and structure of these beds depend on the nature of basement, the volume of sediment load and period of sedimentation (Fig. 2.5).

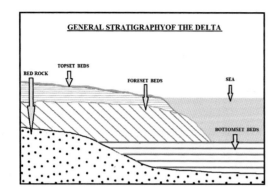

Fig. 2.5 General stratigraphy of the delta

2.6 The Changing Coastline

It appears as the paradox of nature is that in spite of the fact that more than one billion tonnes sediment load carried by the GBM system per year and about 85% of that reaching the coast, the littoral tract in Indian Territory has not grown since the late eighteenth century; on the contrary the sea has encroached inland in many places (Rudra 2012). The comparison of the older maps with recent satellite images reveals such a truth. The rate of encroachment of the coastline varies in different stretches (Bandyopadhyay and Bandyopadhyay 1996). The experts have identified several possible causes of the encroachment of the sea. The major factors governing the erosion of the coast are: (a) sediment flow into the *Swatch of No Ground* (a submarine canyon); (b) frequent attack of the tidal wave; (c) reduction of the downstream flow of the sediment due to dams and reservoirs; (d) sediment trapping in wetlands; (e) auto-compaction of sediment or subsidence due to the earthquakes; The overall mean rate of subsidence in the GBM delta is reportedly 5.6 mm/year (Brown and Nicholls 2015). (f) Depletion of mangroves making the coast open to wave attack and (h) the sea-level rise due to global warming. These factors are working together causing the retrogradation of the western part of coastal Bengal. But the delta has grown appreciably southwards along the Meghna estuary in Bangladesh since it has been the main conduit of sediment influx into the Bay of Bengal (Sarker et al. 2013).

References

Agarwal RP, Mitra DS (1991) Paleogeographic reconstruction of Bengal delta during Quaternary period. In: Vaidyanadhan R (ed) Quaternary deltas of India, memior of the geological society of India 22:13–24

Alam M, Alam MM, Curray JR, Chowdhury MLR, Gani MR (2003) An overview of the sedimentary geology of the Bengal Basin in the regional tectonic framework and basin fill history. Sed Geol 155:179–208

Allison MA, Khan SR, Goodbred SL Jr, Kuehl SA (2003) Stratigraphic evolution of the late Holocene Ganges-Brahmaputra lower delta plain. Sed Geol 155:317–342

Bagchi K (1944) The ganges delta. University of Calcutta, Calcutta

Bagchi K, Mukherjee KN (1979) Diagnostic survey of Rarh Bengal. Department of Geography, University of Calcutta, Kolkata, pp 9–25

Bandyopadhyay S (2007) Evolution of the Ganga Brahmaputra delta: a review. Geogr Rev India 69(3):235–268

Bandyopadhyay S, Bandyopadhyay MK (1996) Retrogradation of the western Ganga-Brahmaputra delta, India and Bangladesh, Possible reasons. In: Tiwari RC (ed) Proceedings of 6th Conference of Indian Institute of Geomorphologists, National Geographer, vol 31, issue 1&2, pp 105–128

Bandyopadhyay S, Das S, Kar NS (2015) Discussion: changing river courses in the western part of the Ganga-Brahmaputra delta by Kalyan Rudra (2014), Geomorphology, 227, 87-100. Geomorphology 250:442–453

Banerjee M, Sen PK (1987) Paleobiology in understanding the change of sea level and coastline in the Bengal Basin during Holocene Period. Indian J Earth Sci 14 (3–4)

Basu SR, Chakraborty S (1972) Some considerations over the decay of the Bhagirathi drainage system. In Bagchi K (ed) The Bhagirathi-Hooghly Basin, Proceedings of Interdisciplinary Symposium, 57–77, 318–321. University of Calcutta, Calcutta

Brown S, Nicholls RJ (2015) Subsidence and human influences in mega deltas: The case of the Ganges-Brahmaputra-Meghna. Sci Total Environ 527–528:362–374

Chakrabarti DK (2001) Archaeological Geography of the Ganga Plain. Permanent Black, Delhi, pp 1–57

Chakraborty SC (1970) Some consideration of the physiographic evolution of Bengal. In: Chatterjee AB, Gupta A, Mukhopadhyay PK, Firma KL (eds) West Bengal. Mukhopadhyay, Kolkata, pp 16–37

References

Chakraborty T, Kar R, Ghosh P, Basu S (2010) Kosi megafan: historical records, geomorphology and the recent avulsion of the Kosi River. Quatern Int 227:143–160

Chatterjee SP (1973) Physiography. In Gazetteer of India/ Country and People, vol I. Govt. of India, p 39

Coleman JM (1969) The Brahmaputra river, channel processes and sedimentation. Sed Geol 3(2 & 3):123–239

Curry JR (1994) Sediment volume and mass beneath the Bay of Bengal. Earth Planet Sci Lett 125:371–383

Curray JR, Emmel FJ, Moore DG (2003) The Bengal fan: morphology, geometry, stratigraphy, history and process. Mar Pet Geol 19:1191–1223

Eaton RM (1996) The rise of islam and the Bengal Frontier, 1204–1760. University of California Press Ltd., London, p 195

Fergusson J (1863) On recent changes in the delta of the Ganges. Q J Geol Soc London 19:321–354

Giosan L, Constantinescu S, Clift PD, Tabrez AR, Danish, M, Inam A (2006) Recent morphodynamics of the Indus delta shore and shelf. Cont Shelf Res 26:1668–1684

Goodbred Jr SL, Kuehl SA (1998) Floodplain processes in Bengal Basin and the storage of the Ganges-Brahmaputra river sediment: an accretion study using ^{137}Cs and ^{210}Pb geochronology. Sed Geol 121:239–258

Goodbred Jr SL, Kuehl SA (2000a) The significance of sediment supply, active tectonism, and eustasy on margin sequence development: late quaternary stratigraphy and evolution of the Ganges-Brahmaputra delta. Sediment Geol 133:227–248

Goodbred Jr SL, Kuehl SA (2000b) Enormous Ganges-Brahmaputra sediment load during strengthened early Holocene monsoon. Geology 28(12):1083–1086

Hirst FC (1915) Report on the Nadia Rivers, Reprinted by Gazetteer Department, Government of West Bengal, Kolkata, pp 3–61

Majumder SC (1941) Rivers of the Bengal Delta. University of Calcutta

Majumder SC (1942) Rivers of Bengal Delta. Reprinted by Gazetteer Department, Government of West Bengal, Kolkata, pp 7–102

Morgan JP, McIntire WG (1959) Quaternary geology of the Bengal basin, East Pakistan and India. Bull Geol Soc Am 70(3):319–342

Mukherjee A, Hazra S (1997) Changing paradigm of petroleum exploration in Bengal Basin. Indian J Geol 69(1):41–64

Mukherjee RK (1938) The changing face of Bengal. University of Calcutta

Oldham (1870) President's address. Proceedings, Asiatic Society of Bengal for February, 1870, Calcutta

Rennell J (1780) A Bengal Atlas (edited by Kalyan Rudra (2016) and published by Sahitya Samsad, Kolkata). London

Rudra Kalyan (2008) Banglar Nadikatha (in Bengali), Sahitya Samsad, Kolkata

Rudra Kalyan (2012) Atlas of changing river courses in West Bengal. Sea Explorers' Institute, Kolkata

Rudra K (1999) The hypothesis of easterly flight of the Ganga water: Fact or Fiction. Indian J Geogr Environ, vol 4, Deptt. of Geography, Vidyasagar University, pp 52–55

Rudra K (2014) Changing river courses in the western part of the Ganga-Brahmaputra delta. Geomorphology 227:87–100

Rudra K (2015) Ref: Changing river courses in the western part of the Ganga–Brahmaputra. Geomorphology 250(2015):454–458

Sarkar MH, Akter J, Rahman M (2013) Century–Scale dynamics of the Bengal delta and future development. In: Proceedings of the 4th International Conference on Water and Flood Management. pp 91–104

Sengupta S (1966) Geological and geophysical studies in western part of Bengal basin, India. Bull Am Assoc Pet Geol 50(5):1001–1017

Sengupta S (1972) Geological framework of the Bhagirathi-Hooghly Basin, In: Bagchi K (ed) The Bhagirathi-Hooghly Basin, Proceedings of Interdisciplinary Symposium, University of Calcutta (Kolkata), pp 3–8

Stanley DJ, Warne AG (1994) Worldwide initiation of Holocene marine deltas by deceleration of sea-level rise. Science 265(5189):228–231

Uddin A, Lundberg N (2004) Miocene sedimentation and subsidence during continent-continent collision, Bengal basin, Bangladesh. Sed Geol 164:131–146

Umitsu M (1993) Late quaternary sedimentary environment in the Ganges delta. Sed Geol 83:177–186

Rivers of the Tarai–Doors and Barind Tract

Abstract

The Mahananda, the Teesta, the Jaldhaka and Torsa along with their numerous tributaries drain the southern slope of the Himalaya and debouch on the plains of Tarai–Doors. While the Mahananda is bifurcated into two branches and ultimately joins the Ganga, three other rivers discharged into the Jamuna in Bangladesh. These rivers are prone to recurrent floods and have produced coalescing fans. One such produced by the Teesta is most conspicuous. The Teesta fan overlies the Barind tract which is also a larger fan formed by post-Pleistocene deposit and stands above the surrounding plains. The rivers like the Kulik, the Karatoya, the Atrayee, the Tangon and the Punarbhava flow southwards through the Barind and have been decayed. The subsidence along the line followed the Jamuna, and compensatory upheaval of the Barind tract have had impacted the recent changes in river courses of this region.

The Ganga and the Brahmaputra enter the plain of Bengal through a 200 km wide passage known as the Rajmahal–Meghalaya gap. The four major tributaries—the Mahananda, the Teesta, the Jaldhaka and the Torsa also flow southward through this gap. The Sankosh flows south along the West Bengal-Assam border and enters Bangladesh to join the Jamuna (Fig. 3.1). The south-west monsoon strikes the Himalaya and generates huge rain as well as the snowmelt water which have caused origin of numerous rivers draining southwards and ultimately discharging either into the Jamuna (Brahmaputra) or the Ganga. The interfluve between the Ganga and the Brahmaputra is drained by an intricate network of rivers which carry a huge sediment load. The rivers debouch on the plains roughly at an elevation of 300 m, becoming wide and sluggish and form a triangular depositional landform called fan. The plains of Tarai and Doors are virtually coalescing fans, and one such produced by the Teesta is most conspicuous. The Teesta fan overlies the Barind tract which is relatively older Pleistocene deposit and stands above the surrounding plains.

The larger rivers like Mahananda, Teesta, Jaldhaka and Torsa have formed overlapping depositional lobes representing multi-dated sediment layers. The discharge hydro-graphs of these rivers are extremely skewed as usual. The shallow cross-sectional areas can hardly accommodate the huge monsoonal discharge, and consequently, floods are an annual event in North Bengal. Where a river drops on the plain, the sudden loss of energy and resultant deposition of the bed leads to a widening and braiding of the channels. The dynamic fluvial systems alter the channel form governed by available discharge, sediment load and declining slope. The drainage map of North Bengal has changed during the known historical period, especially since the

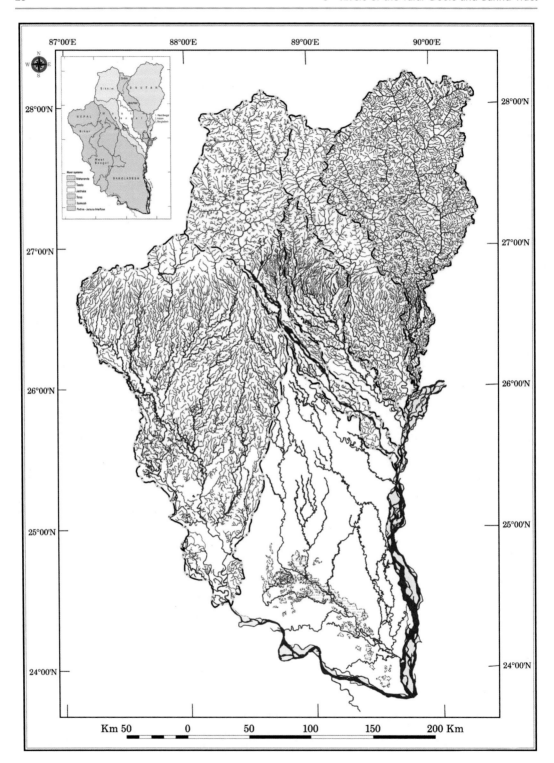

Fig. 3.1 Rivers of Rajmahal–Meghalaya gap. The basins shown here (from the west to the east) are the Mahananda, the Teesta, the Jaldhaka, the Torsa and the Sankosh. While the Mahananda discharges into the Ganga, the four others drain to the Jamuna

second half of the eighteenth century. The rivers of North Bengal have the tendency of avulsion directed towards south-east, and these might be guided by the eastward tilt of the basement (Morgan and McIntire 1959). While all the rivers flow south-eastwards, the Mahananda initially travels south-west and then south-east.

3.1 Mountains to Plains

North Bengal (in West Bengal) has been divided into two well-defined physiographic units—mountains and plains. The 300 m contour line separates these two topographic units (Bagchi and Mukherjee 1983). The hilly region is included in Darjeeling and extreme north-eastern part of Jalpaiguri district. The Himalaya is divided into three parallel ranges, namely Siwalik, Lesser Himalaya and the Greater Himalaya (Valdia 1998). Some major rivers have their sources beyond lesser Himalaya from the snouts of glaciers. The geologists believe that these rivers kept pace with slow upheaval of the Himalaya by the incision of their valleys, so they are called antecedent rivers. The high-level terraces are the evidences of downcutting by the rivers. The rivers are characterised by deep I or V-shaped valleys in mountains and wide-open valleys in plains. The other striking feature is that the rivers change the channel pattern from meandering to braiding as they approach from the mountains to plains. The monsoon rain starts earlier in North Bengal, and the highest rainfall is recorded in the month of July–August contributing to huge discharge in rivers. The minimum flow in rivers is observed in February. Unlike the rivers of South Bengal, here the rivers, except the Mahananda, are replenished by snowmelt water during summer months.

Bagchi and Mukherjee (1983) divided North Bengal into four physiographic units which are geomorphologically different from each other (Table 3.1).

The *Piedmont Zone*, locally known as Doors (doors to the Himalaya), is actually the tilted plains at the base of the Himalaya. It includes entire Siliguri and a narrow strip of Kalimpong subdivision of the Darjeeling district. The Piedmont region is formed due to the coalescing of several alluvial fans within the catchments area of Mahananda, Atrayee, Teesta and Jaldhaka rivers. The zone is enclosed within 600 and 300 m contours, respectively. This area is the transitional belt between Himalaya and the floodplains of North Bengal. The *Active zone* includes Jalpaiguri, Alipurduar, Koch Bihar and Uttar Dinajpur covering 5979 km^2. Since the gradient is gentle, the rivers flow in meandering courses and floods are common phenomena during the monsoon. The suspended loads are carried away from river channels by floodwater while larger gravels and pebbles are left behind in the beds. The *Mature zone* is composed of lateritic or older alluvial humps. This old mature plain locally known as *Barendrabhumi* looks like a mega fan covering 1620 km^2 within the districts of Uttar Dinajpur, Dakshin Dinajpur and Maldah. The Teesta fan is overlaid and juxtaposed with the Barind tract in Bangladesh (Chakraborty et al. 2010). The Barind is also treated as the area of compensatory upheaval linked with the subsidence of central part of the Bengal basin (Morgan and McIntire 1959). The rivers like Karatoya, Atrayee, Punarbhaba have flowed southward through Barind. The *Moribund zone* is the plain covering an area of 4570 km^2 which is enclosed by 27 m contour and the natural levees of the Ganga, Mahananda and Atrayee. This physiographic unit is further

Table 3.1 Geomorphic units of North Bengal

Geomorphic units	Area (km^2)
Piedmont zone (Tarai and Doors)	7282
Active zone	5979
Matured zone	1621
Moribund zone	4542

divided into two subzones, the *Diara* and the *Tal*. The *Diara* is relatively flat land formed by the deposition of younger alluvium. In the west, most of the land below 27 m is recognized as the *Diara*. It is a boggy area with pockets of marshy lands. The *Tal* is relatively higher lands adjoining the *Diara*, and generally formed by older alluvium.

3.2 Rivers of the Himalayan Front

The tributaries of the Ganga and the Brahmaputra drain southern slope of the Himalaya and enter Bengal through the Rajmahal–Meghalaya gap. The Teesta, the Jaldhaka, the Torsa and the Sankosh are the main tributaries to the Brahmaputra or Jamuna in Bangladesh. The Mahananda and its principal distributary, the Fulohar join the Ganga at Godagari ghat of Bangladesh and Manikchak ghat at Maldah, respectively. The rivers of North Bengal have a common geopolitical characteristic. All major rivers, except the Mahananda, originate beyond the northern boundary of West Bengal, and these rivers including Mahananda flow southward through Bangladesh in their lower reach (Govt. of WB 2001) (Table 3.2).

3.3 Characteristics of North Bengal Rivers

- All major rivers of North Bengal are shared by two or more States, and they are called transboundary rivers. They have their origin in Bhutan, Sikkim, Darjeeling and Tibet and flow southwards through West Bengal and ultimately cross the Indo-Bangladesh border discharging either into the Ganga or the Brahmaputra.
- The Himalayan Rivers debouch on the plain of North Bengal, where the rivers become suddenly sluggish and deposit substantial sediment load, forming a triangular feature known as the fan. The North Bengal plain as a whole is a series of overlapping fans extending southward beyond the Indo-Bangladesh border.
- The declining slope and increasing deposition on the riverbed have formed braided channels in North Bengal, and the intricate network of interlacing channels is characteristic geomorphological features found in cases of all the major rivers. The point bars are dynamic and constantly change their shape and geometry, especially in the monsoon months.
- The rivers of North Bengal have the tendency to change their courses constantly.

The geology as well as the physiography of the basin have their impact on the river courses. When a river flows through the steep gorge in the Himalaya, the rate of downcutting is greater than the lateral cutting, but when it reaches the plain, the rate of downcutting declines and lateral erosion increases. Thus, the river widens its valley.

- All the rivers of North Bengal are multi-channelled conduits transferring water and sediment load. When the river is bank-full during the peak of the monsoon all

Table 3.2 River basins of North Bengal

Major drainage basin	Total catchment area (km^2)	Area in West Bengal (km^2)	Annual discharge (MCM)	Total suspended sediment load (million tonnes)
Teesta	12,370	3294	29,947	14
Jaldhaka	5792	3621	17,212	4
Torsa	7914	3375	23,097	11
Mahananda	14,946	9802	23,129	1
Fulohar	12,707	414	26,260	13

3.3 Characteristics of North Bengal Rivers

Table 3.3 Estimated flow of the Teesta at Chilmari

Basin area 12,370 km²	Estimated flow leaving the basin (MCM)											
	Jan	Feb	Mar	Apr	May	June	July	Aug	Sept	Oct	Nov	Dec
Chilmari (outfall)	603	754	1176	1742	3025	4591	4780	5250	3471	3160	863	533

minor interlacing channels are combined together to form a single channel but when the discharge hydro-graph falls below the threshold limit the minor channels reappear within the cross-sectional area of the river.

- The rivers of North Bengal except Mahananda have a tendency to flow south-eastward. In fact, the interfluve between the Teesta and the Mahananda acts as the divide separating the two river systems.
- The heavy monsoon rainfall, shallow cross-sectional area, declining slopes, huge sediment load are the major causes to make the North Bengal flood prone. The North Bengal Flood Control Commission and State Irrigation Department relied on embankments as the flood control measures, but those embankments failed to ensure protection during many devastating floods. The long-term effects of jacketing the river appear to be detrimental as it caused rapid deposition in the riverbed and diminution of the water-holding capacity (Mahalanobis 1927).
- The east–west aligned railway and highways have an immense impact on hydro-geomorphology of North Bengal. The alignment of these communication lines with inadequate narrow culverts created serious problem of drainage congestion and an expansion of the flood contour (Fig. 3.2). The devastating flood of 1922 was ascribed to the intense rain as well as expansion of road and railway with inadequate culverts (Saha 1933; Ray 1932). Since the passages of water are artificially constricted at the railway-road bridges, the rivers have the tendency to adjust with these structural interventions either by widening their valleys or by altering the courses in upstream and downstream sections.

The Brahmaputra originates from Manas Sarovar in the Tibetan Himalaya and flows eastward. It is known as Tsangpo in Tibet and Brahmaputra in Assam and Jamuna in Bangladesh. The Brahmaputra ultimately discharges into the Bay of Bengal. The total length of river is 2900 km, and its basin area covers 552,000 km². The Brahmaputra has taken a southward bend near a peak called Namchabarwa and is described as the Dehong while crossing the Himalaya. It takes a hairpin bend while entering Assam and flows westward for a length of about a 600 km before it reaches the Indo-Bangladesh border. The river turns southward near Dhubri and enters Bangladesh as the Jamuna. The Ganga and the Brahmaputra were originally two independent rivers with separate outfalls (Rennell 1780). The Brahmaputra, prior to 1830, flowed through Mymensingh and Syllet basin. It subsequently shifted westward and started to flow southward to join Padma at Goalando. The Jamuna which was formerly an offshoot of the Brahmaputra was gradually enlarged and continued to carry the bulk of water. This caused decay of the Mymensingh outlet. Many rivers join the Brahmaputra in the Indian Territory. It receives the tributaries like the Sankosh, the Torsa, the Jaldhaka and the Teesta. The Brahmaputra has formed a unique and braided channel network within its valley. These systems are very dynamic and change continuously.

3.4 The Teesta and Its Tributaries

The Teesta is the most significant river of Sikkim and North Bengal. It originates from Kangtse glacier of North Sikkim at an altitude of 6200 m and flows through Sikkim, West Bengal and Bangladesh and finally falls into the Jamuna or Brahmaputra near Chilmari of Bangladesh (Fig. 3.3). The length of the river is 414 km, and it flows about 172 km through the mountains of Sikkim before debouching on the alluvial plains

Fig. 3.2 Impact of a bridge on the course of the Jaldhaka

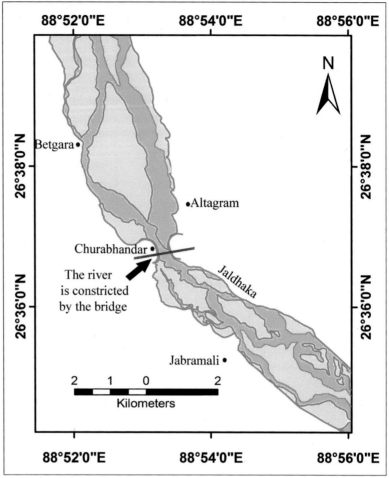

of the Himalayan front. The Teesta flows 123 km through Indian plains before it crosses the Indo-Bangladesh border. It flows for about 121 km in Bangladesh. The catchment area of this river covers 12,370 km². The name Teesta was derived from *Trisrota,* meaning three streams which are the Karatoya, the Attayee and the Purnabhaba.

The two tributaries, namely Lachung and Lachen meet each other near Chungthang of North Sikkim. The combined flow of Lachung and Lachen, after emerging from Chungthang valley, goes by the name of Teesta River. It receives many tributaries during its southward journey, namely Lish, Ghish, Chel, Neora, Rangpo and Rangit (Table 3.4). After flowing for a long distance through the mountains, the river enters into the plain of West Bengal at Sevoke near Siliguri. It fans out from this point, and the floodplain widens to about 4–5 km. at places. It is joined by Dharlariver near Domohoni and Karalariver near Jalpaiguri. These streams carry huge amount of sediments from their upper catchment. The average suspended load carried by the Teesta is 7.71 mg/litre. It was further estimated that the river carries 14 million tonnes of suspended load annually. A high water level in the Teesta often retards the speedy outflow from its tributaries, resulting in drainage congestion. This stream, while flowing through Jalpaiguri, receives storm and sewage water of the town. Of late owing to sedimentation at the outfall point, the discharging capacity of Karala has been impaired Table 3.3 describes mean monthly flows of Teesta at

3.4 The Teesta and Its Tributaries

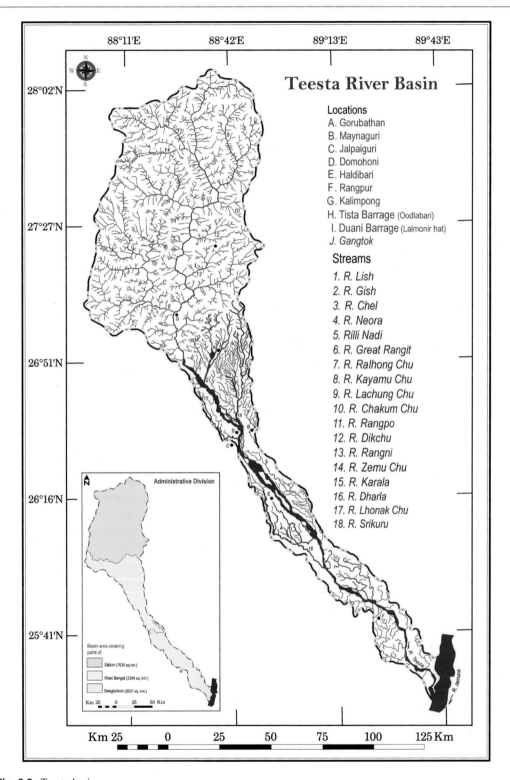

Fig. 3.3 Teesta basin

Table 3.4 Important tributaries of the Teesta

Right bank tributaries	Origin	Length (in km)	Catchment area (in km^2)
Lachung	Donkya Rhi	58	820
Bakchu-Dikchu	Chola ridge	32	250
Rongni	Yamgmenkhang ridge	34	250
Rangpo	Sikkim-Tibet Border	48	550
Rangit	Janre-Danra hills	80	2000
Karala	Baikunthapur Forest	70	140
Left bank tributaries			
Lachen	Ranchung Chho	80	1720
Lish	Combination of Lish Khola and Turung Khola		64
Gish	Git Khola	35	160
Chel	Pankhasari reserve forest	54	390
Neora	Rechilla Chawk	58	275
Relli	Khampong reserved forest	32	165

Source Master plan for flood management and erosion control in North Bengal (2001)

Chilmari in Bangladesh. It appears that about 60% of the annual flow passes during four monsoon months.

3.5 The Teesta Fan

The Teesta has bifurcated the foothills of Eastern Himalaya into two parts, the Terai in the west and Doors in the east. The fans have developed along Terai and Doors area located along the frontal plains of the Himalaya. The Teesta itself formed a huge fan at the foothills of Himalaya, extending down to Bangladesh. It is bordered by Mahananda in the west and the present course of Teesta in the East, covering 18,000 km^2. The Teesta Fan is a triangular geomorphic unit characterised by the presence of old and modern drainage line. Width and length of the mega fan are 145 and 166 km, respectively. The surface of the fan is convex (see Fig. 2.4). While this mighty fan is flanked by the Teesta and the Mahananda, a large number of streams traverse it (Chakraborty and Ghosh 2010). The Teesta Fan has a long depositional history which initiated from the Pleistocene period. The post-Pleistocene retreat of glaciers resulting an increase in sediment load and discharge in all Himalayan rivers are supposed to be the major factors contributing to the formation of this mega fan.

When a river debouches at the junction of hills and plains, it becomes unconfined and immediately starts to spill. The velocity decreases with the fall of gradient. There were multiple phases of deposition during formation of the fan. The Teesta along with its tributaries and distributaries continuously supplies sediments to the fan. The flash flood during monsoon often deposits the large amount of detrital sediments brought from the mountain, and deposition continues southwards from the apex of the fan. Sometimes braided channels are unable to carry sediments, and they get deposited to form channel bar or point bar. These sediments are mainly fine-grained silts and sands and such sedimentation rarely contributes as fan materials until the river overtops the bank or the river shifts its course. Mainly the fan materials carried by the Teesta are coarse-grained, poorly sorted and immature. The comparatively younger northern part of the fan overlies the older southern part. The sediment compositions in these two hubs are different. In the upper or northern reach, boulders and gravels are predominating and sand, silt and clay are found in less amount. The sediments size is quite large, but in the

3.5 The Teesta Fan

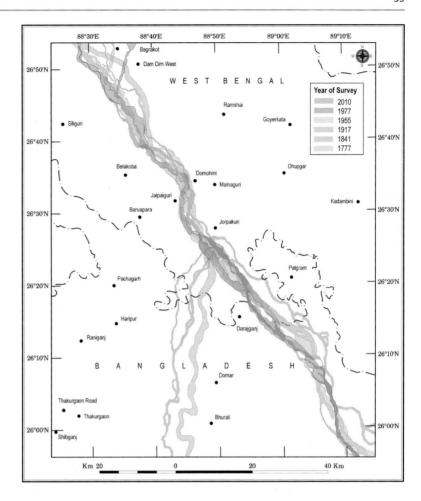

Fig. 3.4 Map showing avulsion of the Teesta

southern part, the size of the materials is comparatively smaller. Sediment size tends to decrease from the mountain front to the distant place. The roundness of the pebbles increases with distance from the apex of the fan.

The shape and topographic expression of the Barind tract (*Barendrabhumi*) typically resembles a mighty fan which was formed by the Pleistocene and Holocene sediment after the recession of the sea level during Pleistocene glaciations (Umitsu 1993). This is a unique geomorphic feature covering 9324 km^2, which is flanked by Mahananda in the west and Karatoya (a distributary of the Teesta) in the east (Morgan and McIntire 1959). It is reported by the geologists that the sea level was lowered about 120 m during the Pleistocene when Himalayan glaciers advanced southwards and the discharge in rivers diminished appreciably. But post-Pleistocene melting of the glaciers contributed a huge snowmelt water and sediment load in rivers.

These sediment loads were ultimately deposited to form a triangular geomorphic unit and that is known as the Barind Tract. Some beheaded streams like Punarbhaba, Atrayee, Karatoya have cut across the Barind or Pleistocene deposits to maintain their southward course. These rivers were beheaded when the Teesta migrated eastwards (Fig. 3.4) and now replenished by the monsoon rain and base flow from groundwater pool. The Teesta fan is characterised by a divergent drainage pattern radiating southward from Jalpaiguri. The rivers which traverse the fan are comparatively narrow, more sinuous and entrenched (Fig. 3.5).

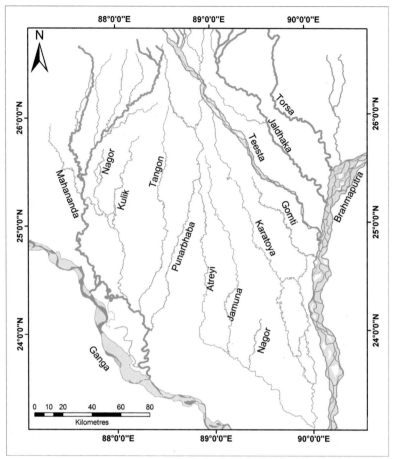

Fig. 3.5 Rivers of the Barind tract

(1) Since the easterly flight of the Teesta, minor distributaries, namely Karatoya, Atrayee, Punarbhaba and Tangon were beheaded and gradually decayed, resulting in the diminution of sediment supply to the fan. Multi-lobate geometry is the major feature of the Teesta fan. Three distinct depositional lobes have been identified which overlap with each other. The formation of these overlapping lobes has been attributed to the avulsion of river, fluctuation of the discharge and rapid sedimentation. The mode of deposition within the fan varies in different parts. In Doors and Terai, stratigraphy is distinctly exposed along the shelving riverbanks, whereas in the southern part the stratigraphical sequence is disturbed by the continuous cultivation of land by the farmer.

3.6 Human Intervention and River Style

The course of the Teesta below Sevok is distinctly different than its upper reach. The bed load dropped due to sudden decline in slope has created formidable obstruction to river channel and thereby compelled it to split into several channels. Linear embankments were built along the bank to combat flood. But the river often undermines the bank and leads to the collapse of overlying embankments. The river has adjusted itself subsequent to the construction of a mighty barrage at Gajoldoba in 1997. This can be distinctly shown by comparing the course of river in 1972 and that of 2010 (Figs. 3.6 and 3.7).

3.6 Human Intervention and River Style

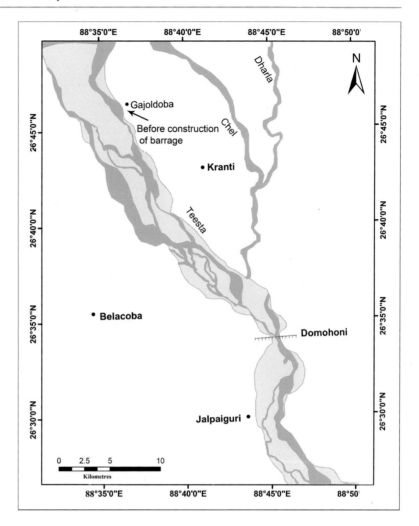

Fig. 3.6 Course of Teesta before construction of the Gajoldoba barrage (1972)

The river has been strikingly constricted at Gajoldoba site since the barrage was constructed. Similar constriction has also been observed at the places where railway or highway crosses the river. In all cases of constriction, the river has become wide in both the upstream and downstream sections. In fact, the bridges or roads with inadequate passage of water under them create drainage congestion in the upstream section, especially during a flood. The bridge virtually creates a steeper hydraulic gradient between the upstream and downstream stretch and the resultant impact is felt in both the stretches.

3.7 The Jaldhaka System

The river Jaldhaka originates from a lake in Sikkim Himalaya at an altitude of 4400 m. It flows through Sikkim, Bhutan, West Bengal and Bangladesh. The length of the river up to Indo-Bangladesh border is 209 km. It flows about 52 km through Rongpur of Bangladesh and joins Jamuna (Fig. 3.8). The total catchment area is 5792 km^2, and 50% of the catchment is mountainous and the remaining part is alluvial plain. The volume of water flowing through this

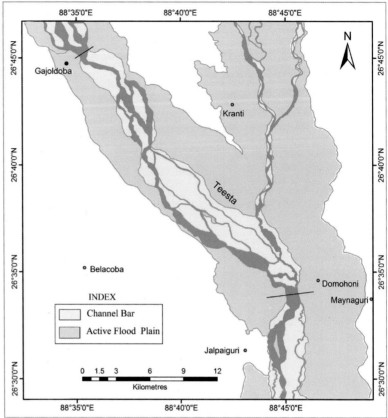

Fig. 3.7 Course of Teesta (2010) and its active flood plain after construction of the barrage

river between January and October is estimated as 17301 MCM, and nearly 83% of the annual flow is generated during rainy season (Table 3.5). The annual suspended sediment load is 4.2 million tonnes. Many mid-stream bars compel the river to flow through braided channels. The river is also known as 'Mansai' in its lower reach.

The comparison of survey of India's topographical sheets (1936) with recent Landsat satellite image (2016) reveals that the river has changed its course at some stretches. Like that of the Teesta, the channel is strikingly constricted at places where the railway and highway have been aligned over the river and it has widened in both upstream and downstream of the bridges. The river has changed its course appreciably through the process of channel migration in some stretches, and the changes are shown through multi-dated maps.

3.8 The Tributaries

A number of tributaries have joined the river Jaldhaka from left. An important tributary named Assom khola joins the river from the left bank and further down it receives another river called Nichu from the right side where it forms the boundary between Bhutan and India. Further downstream, it receives another tributary named Bindukhola from left and takes the name Jaldhaka. When the river enters in the Indian Territory, it is joined by Jiti stream. Another stream called Jhalong khola joins and drains the interfluves between Bindu khola and Jiti. Below the confluence of Jiti, Jaldhaka is bifurcated into two channels and suddenly turns at right angle towards the south and again splits into two channels, namely Hatinala and Jaldhaka. These two channels are subsequently combined and

3.8 The Tributaries

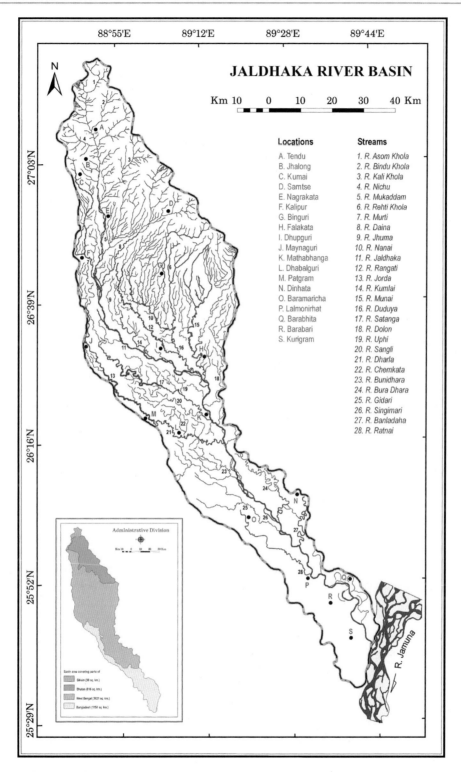

Fig. 3.8 Jaldhaka River basin

Table 3.5 Estimated flow of the Jaldhaka River (in MCM)

Basin area 5792 km²	Months											
	Jan	Feb	Mar	Apr	May	June	July	Aug	Sep	Oct	Nov	Dec
Kurigram (outfall)	46	41	72	532	1622	3906	4469	3280	2625	708	−31	−59

NB. Since evapotranspiration and infiltration exceed rainfall and over exploitation of groundwater leads to diminution of effluent seepage, the flow in the river appears negative in November and December

continue to flow towards the south. There is a railway bridge over the Jaldhaka situated downstream of the northern border of the Jalpaiguri district where three tributaries Ghatia, Diana and Murti come from the north-east. Below the confluence with Diana, the river flows south through more than one channel. Near Betgara, the river turns to the left and flows in a south-easterly direction. Here, it receives 'Duduya' and 'Mujnai' from the left. From the confluence of Mujnai, the Jaldhaka suddenly takes a turn at right angle and flows south for about 5 km near the town of Mathabhanga, where the river becomes wider and flows through two channels which are again united at Mathabhanga. A tributary named Sutanga joins Jaldhaka from right side. Subsequently, Bara meets Jaldhaka from the west and Girdhari from east. The river becomes wider as it flows south through the plains where it changes its geometry of meandering (Figs. 3.9 and 3.10).

3.9 The Torsa System

The Torsa is the third largest river in the North Bengal after Mahananda and Teesta in terms of catchment area. It originates from Chimbay valley in Southern Tibet at an altitude of 7065 m where it is known as Machu. It flows through Bhutan where it is known as Amo chu. The mountain streams, namely Kangpo and Kylang are combined to form Amo chu in Bhutan Himalaya. The river receives three important right bank tributaries, namely Chimkiphu, Tangka and Namchu khola and five other left tributaries, namely Tromo, Yak, Ripolo khola, Netap khola and Pa. The Amo chu enters West Bengal near Hasimara border and takes the name Torsa.

It flows about 71 km in Tibet, 80 km in Bhutan, and 99 km in West Bengal and 45 km in Bangladesh (Fig. 3.11). The total length of the river is 295 km, and the catchment area including Bangladesh covers 7914 km². The annual volume of water flowing through this river is 23,097 million cubic metre (Table 3.6). The Torsa carries 11 million tonnes of sediment load/year. When river flows through Jalpaiguri and Koch Bihar, it becomes wide and shallow. Declining slope and increasing volume of water compel the river to throw spill channels which facilitate discharge of excess water during high flood or in rainy season. After debouching on the plain, the cross-sectional area of the river is so reduced that it cannot accommodate the peak discharge of monsoon and consequently flooding is a recurrent hydro-geomorphological phenomenon in the lower catchment.

It flows southward up to Koch Bihar and then moves south-eastward to join Jamuna in Bangladesh. In the lower reach, Holong is an important right bank tributary to the Torsa and the important left bank tributaries are Kaljani, Napania and Gadadhar. While traversing through the plains, it throws off a spill channel named Sil Torsa near Hasimara Bridge. The Sil Torsa rejoins the Torsa at a point where the National Highway crosses the river. Closer to the National Highway no. 31 near Madarihat (Alipurduar district), the Torsa receives another tributary called Choto Torsa.

There is another spill channel which takes off from Choto Torsa and joins the Jaldhaka below Falakata. Below Kulpani, the Torsa throws off two distributaries called Dharla and Mansai, and both flow to Jaldhaka. In the early decades of the twentieth century, both these two distributaries carried the bulk of the discharges but now have

3.9 The Torsa System

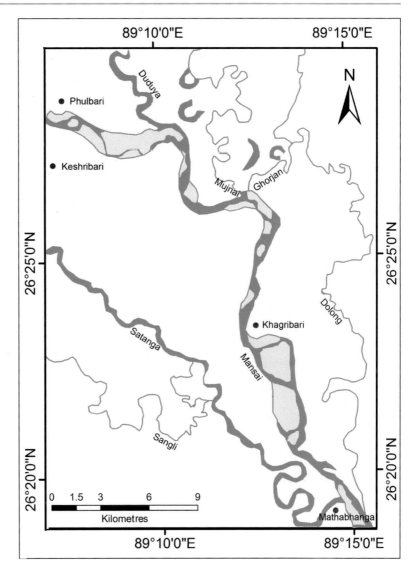

Fig. 3.9 Jaldhaka in 1936

gone dry. The Raidak originates from the Bhutan Himalaya and throws off three distributaries, namely Dhaula, Raidak-I and Raidak-II. The Dhaula and Raidak I are subsequently combined together and join Torsa while Raidak-II flows south-east to join Sankosh.

3.10 The Mahananda River System

The Mahananda is the only important tributary to the Ganga that drains an extensive part of North Bengal. It originates from Mahalidram of Darjeeling Himalaya and debouches on the plains near Siliguri. It flows south-westward into Bihar and is bifurcated into two branches at Benibari (near Bagdob); the western branch is called the Fulohar and carries the bulk of the discharge and joins the Ganga near Manikchak ghat in Maldah (Fig. 3.13). The other branch carries the name 'Mahananda' but is virtually disconnected from its feeder and does not receive any upstream flow except during the high flood. The water contributed by the Balason and other Himalayan rivers now flows through the Fulohar branch. The eastern branch or the Mahananda receives Nagor,

Fig. 3.10 Jaldhaka in 2016

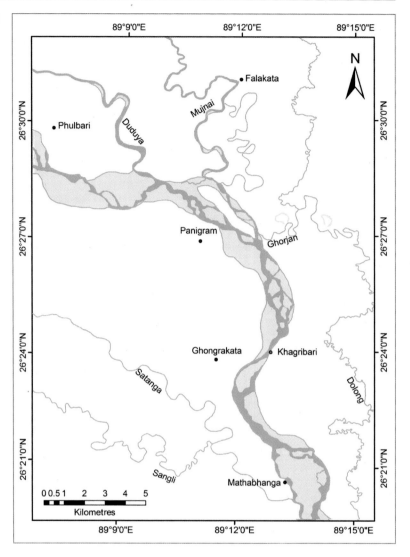

Kulik, Chiramati, Tangon and Punarbhaba as the left bank tributaries (Fig. 3.13). A branch of Ganga named Kalindri carries excess floodwater and joins the Mahananda at Nimsorai Ghat close to Maldah town. This channel had been the principal outlet of the Ganga water till fourteenth century (Mukherjee 1938). When Kalindri carried the principal freshet of the Ganga, the lower Mahananda below Nimsorai ghat was a part of it. It has been estimated that the Fulohar carries maximum discharge in the month of July (7238 mcm). The river goes dry in the months of November and December because of minimum rainfall, excessive evapotranspiration and the diminution of base flow from the groundwater pool. The eastern branch of Mahananda having a comparatively smaller catchment (9627 km^2) carries much less water than the Fulohar. The maximum flow is generally observed in the month of July (5385 mcm), while the negative flow is estimated in the months of November and December (Table 3.7). The westerly flight of flow or avulsion from the Mahananda to the Fulohar has been a recent phenomenon and unlike the Teesta it was not a sudden event during a single flood rather the process took about three decades commencing from the late 1980s (see Figs. 3.12 and 3.13).

3.10 The Mahananda River System

Fig. 3.11 Torsa River basin

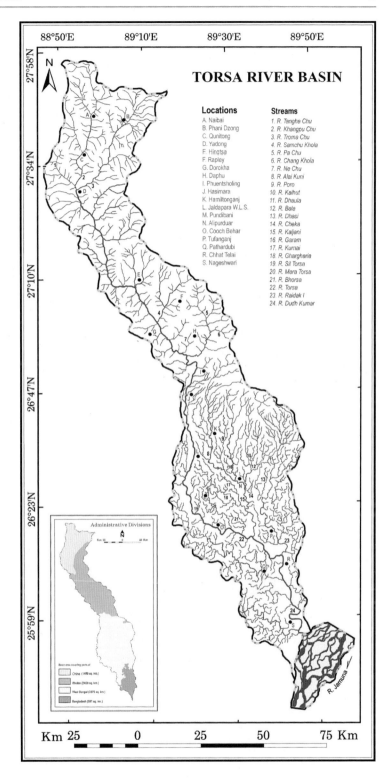

Table 3.6 Estimated flow in the Torsa basin

Basin area 7914 km²	Estimated runoff leaving basin (mcm)											
	Jan	Feb	Mar	Apr	May	June	July	Aug	Sep	Oct	Nov	Dec
Nageshwari (outfall)	145	120	181	747	2106	4872	5499	4469	3330	1395	170	64

Table 3.7 Estimated flow of the Fulohar and the Mahananda (in MCM)

Basin	Basin area (km²)	Months											
		Jan	Feb	Mar	Apr	May	June	July	Aug	Sep	Oct	Nov	Dec
Fulohar (Manikchak)	12707	195	151	117	642	2039	5088	7238	5783	4339	961	−133	−159
Mahananda (Godagarighat)	9627	590	503	367	733	2189	3887	5385	4638	4077	1085	−142	−183

3.10 The Mahananda River System

Fig. 3.12 Tributaries of Mahananda

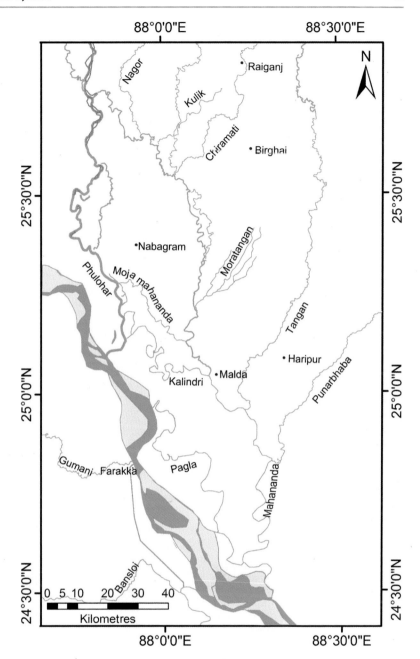

The lower reaches of both the Mahananda and Fulohar are extremely fertile, and irrigation therein is either river-lift or based on groundwater. These two factors are combined to make the rivers fordable in the winter season.

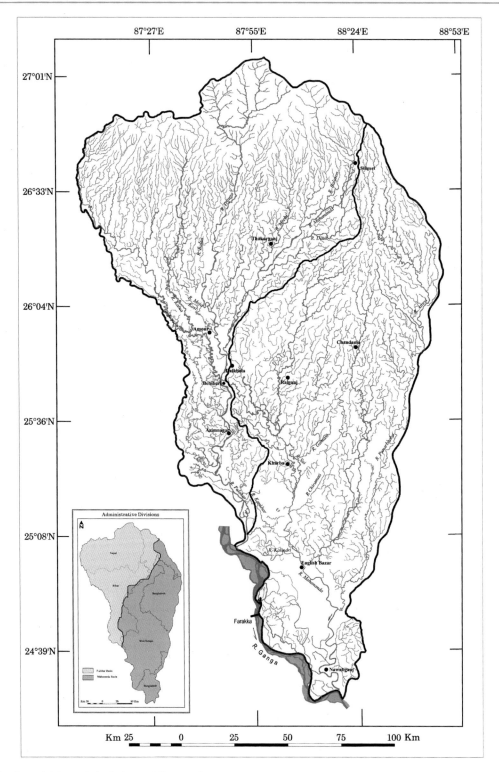

Fig. 3.13 Mahananda basin is now bifurcated into two sub-basins. The western one is Fulohar, and the eastern one is the Mahananda

References

Bagchi K, Mukherjee KN (1983) Diagnostic survey of West Bengal (North). Deptt. of Geography, University of Calcutta

Chakraborty T, Ghosh P (2010) 'The Geomorphology and sedimentlogy of the Tista Megafan, Darjeeling Himalaya: implications for Megafan building Processes. Geomporphology 115:252–266

Chakraborty T, Kar R, Ghosh P, Basu S (2010) Kosi Megafan: historical records, geomorphology and the recent avulsion of the Kosi river. Quatern Int 227:143–160

Govt. of West Bengal (2001) Master plan for flood management and erosion control in North Bengal, Phase-I. Draft Final Report, vol I

Mahalanobis PC (1927) Rainfall and floods in North Bengal, 1870–1922, Department of Irrigation, Bengal Government

Morgan JP, McIntire WG (1959) Quaternary geology of the Bengal basin, East Pakistan and India. Bull Geol Soc Am 70(3):319–342

Mukherjee RK (1938) The changing face of Bengal. University of Calcutta

Ray PC (1932) Life and experience of a Bengali Chemist, II, 159–160, Reprinted by Asiatic Society (1996), Kolkata

Rennell J (1780) A Bengal Atlas (edited by Kalyan Rudra and reprinted by Sahitya Samsad (2016)). London

Saha MN (1933) Need for a hydraulic research laboratory, Reprinted (1987) in *Collected Works of Meghnad Saha*, Orient Longman, Calcutta

Umitsu M (1993) Late quaternary sedimentary environment and landform evolution in the Bengal Lowland. Sed Geol 83:177–186

Valdia KS (1998) Dynamic Himalaya, Orient Longman

The Dynamic Ganga

4

Abstract

The Ganga flows through Bengal for about 457 km between the Rajmahal hills and the Bay of Bengal. It receives the Jamuna and Meghna as left bank tributaries while flowing through Bangladesh and throws off the Bhagirathi, the Bhairab–Jalangi, the Mathabhanga–Churni, the Garai and the Tentulia as right bank distributaries. The left bank distributaries like the Kalindri, the *Chhoto* Bhagirathi, the Pagla and the Boral have gone dry. The Ganga continuously changes its course through avulsion and meander migration. It erodes fertile land and displaces many villages every year. The emergence and shifting of the capital towns of Bengal happened also due to changing courses of the Ganga and its distributaries. The Holocene tilt of the delta governed the eastward flow of the Ganga, and consequently many distributaries have decayed. The erosion of the bank, especially in Malda district of West Bengal, has been accelerated after the commissioning of the Farakka barrage.

The Ganga along with its tributaries and distributaries constitutes the largest fluvial system in India. Its basin area covers 1.08 million km^2 and spreads over four countries which are India, Nepal, China and Bangladesh (Fig. 4.1). The Ganga basin extends over eleven states of India covering 26% of its territory. The river enters West Bengal along Rajmahal hill of Jharkhand and flows for about 120 km to reach Mithipur (near Jangipur, Murshidabad district) where it is divided into two major distributaries–the Ganga (called Padma in Bangladesh) itself and the Bhagirathi. The Ganga/Padma carries the bulk of flow and delineates the Indo-Bangladesh border for about 90 km from Mithipur to Jalangi, which is the last Indian settlement and a border town. The Ganga flows 132 km eastwards through the territory of Bangladesh and receives the Brahmaputra or Jamuna as a left bank tributary at Goalundo. The combined channel (hereafter known as the Padma) flows for 115 km south-eastwards to join the Meghna at Chandpur and finally discharges into the Bay of Bengal. In between the Rajmahal and the Farakka barrage, the three moribund distributaries of the Ganga, namely Kalindri, Pagla and *Chhoto* Bhagirathi (a repetition of the name given to the western branch of the Ganga which empties itself into the Bay of Bengal at Gangasagar), take off from the left bank in Maldah district and rejoin with the parent river directly or indirectly further downstream. While the first one (the Kalindri) joins the Mahananda, the two other channels (the Pagla and the *Chhoto* Bhagirathi) unite again before joining the Ganga (Fig. 4.7). The Boral, which is now disconnected from the Ganga, took off at Charghat and flowed eastwards to join the

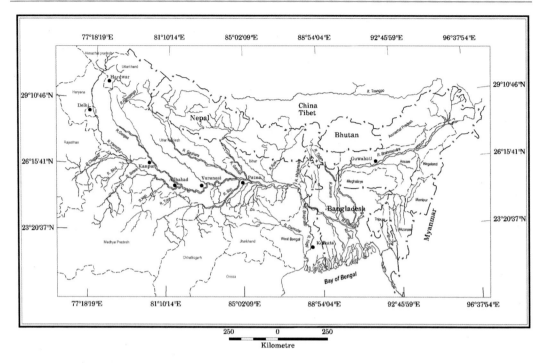

Fig. 4.1 Rivers Ganga–Brahmaputra basin

Jamuna. The Bhagirathi, the Bhairab–Jalangi and Mathabhanga are three important right bank distributaries. The first two unite at Mayapur–Nabadwip and flows southwards as the Hugli towards the sea. The Mathabhanga is bifurcated into two branches, Churni and Ichhamati, at a place called Majhdia; while the former flows westwards to join the Bhagirathi at Shibpur Ghat near Payaradanga, the latter flows southwards and takes the name Kalindi in Sundarban before discharging into the sea (Fig. 4.2).

Bengal was politically divided in 1947; while dividing the country on the basis religion, the international boundary was drawn across 54 rivers which enter Bangladesh from India. The Ganga was chosen as international border between Rajshahi district of Bangladesh (erstwhile East Pakistan) and Murshidabad district of India. Since then the Ganga has changed its course and flows far away from the border.

The eastward tilt of the delta has enabled the Ganga/Padma to incise its valley deeply during the late Holocene period and that resulted beheading of the many distributaries which earlier flowed southwards and independently discharged into the Bay of Bengal. The southern littoral tracts of Bengal, popularly known as the Sundarban, are drained by the interlacing network of numerous tidal creeks. Most of these creeks do not have any upstream water supply and are fed by tidally induced water. These creeks are also treated as a part of the Ganga system and account for the name *Satamukhi* (one with a hundred mouths) used in the mediaeval Bengali literature to describe the Ganga. The six major creeks draining the Indian Sundarban are Saptamukhi, Thakuran, Matla, Bangaduni, Gosaba and Haribhanga. The Bartala or Muriganga which flows between Sagar Island in west and Namkhana–Kakdwip in the east is the only channel in West Bengal rendering freshwater to the Sundarban. The major creeks flowing through the Bangladesh are Malancha, Pusur, Selagang, Baleswar, Buriswar, Patuakhali and Meghna.

Fig. 4.2 Mathabhanga, Churni and the Ichhamati

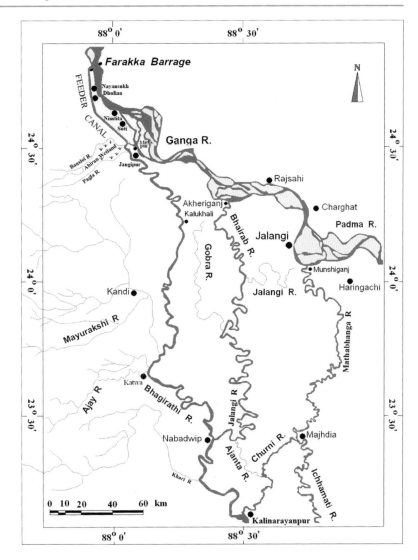

4.1 The Ganga through West Bengal

The Ganga with its tributaries and distributaries constitutes a complex drainage network draining Bengal. The river changes its course frequently, and such change is governed by impinging flow and unconsolidated bank stratigraphy. The Farakka barrage was built across the Ganga to impound and divert water, but it has also trapped the sediment load and led to the formation a large bend between the Rajmahal hill and the Farakka barrage. While the river erodes its left bank in the stretch between the Rajmahal and the Farakka barrage, the opposite bank is severely affected in the downstream stretch. In the remote past, the Ganga flowed through different paths adopting a series of distributaries and those channels have now decayed. The ruins of capital towns like Gaur, Tanda and Pandua now stand far away from the Ganga. The present channel has a tendency to migrate laterally. The research on multi-dated maps reveals how the Ganga has altered its course since the mid-nineteenth century. The changing courses of the Ganga in both upstream and downstream of the Farakka barrage have caused many problems like erosion of the fertile land from one bank, emergence of new land along opposite bank, population displacement and Indo-Bangladesh border dispute.

The Ganga oscillates laterally within its meander belt, and old courses are often left behind on the floodplain. It is understood that in the process of meander migration, the Ganga may go back to its former course, but there is no fixed time of such reversal of courses. The tributaries draining from the Himalaya often avulse from one side to other. This is common in the Ganga–Brahmaputra plain. The avulsion of the Kosi in 2008 is an example of the migration of rivers in the Himalayan foothill where the valley is not enclosed by the high mountains (Fig. 4.3). Notably both the Ganga in Bengal and Kosi in Bihar were tamed for different purposes. The former was intercepted at Farakka to divert water and resuscitate the port of Kolkata, and the latter was embanked for the flood control. In both cases, sediment loads were trapped either between the embankments or in the bed upstream of the barrage. The cross-sectional area of the Ganga has been reduced due to rapid sedimentation in

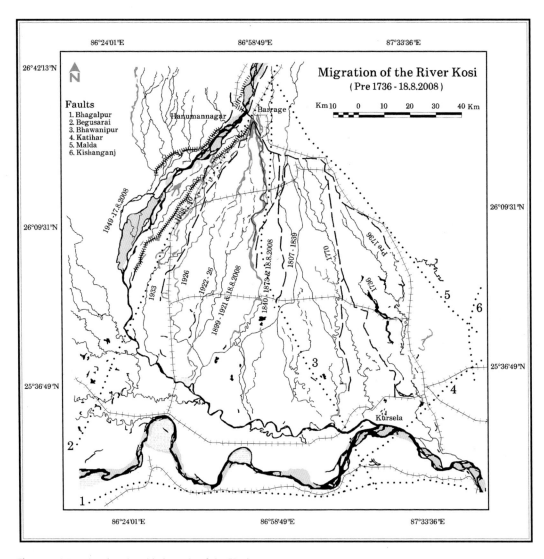

Fig. 4.3 Map showing the old channels of the Kosi

between Rajmahal and Farakka compelling to form an eastward bend (Rudra 2006). The flood control embankments were built along the bank of the Kosi since 1955, and a barrage was built at Bhimnagar (in Nepal) in 1963 to ensure both flood moderation and irrigation. Notably the barrage has no storage capacity and can only induce water into the irrigation canal. Still it is often wrongly described as a flood moderator. However, the river bed above Bhimnagar and downstream recorded accelerated deposition and gradually went high above adjoining floodplains and ultimately caused avulsion on 18 August 2008 (Sinha 2008; Mishra 2008; Rudra 2009). The comparison of multi-dated maps reveals the Ganga has changed its course laterally but it has migrated eastward unilaterally since the Farakka barrage was commissioned in 1975. Since the mighty structure was built across the Ganga and impeded flow of water and sediment load, the river impinged the left bank upstream of the barrage and encroached the opposite bank in the downstream (Figs. 4.4 and 4.5). The other factors facilitating recurrent bank failure are the presence

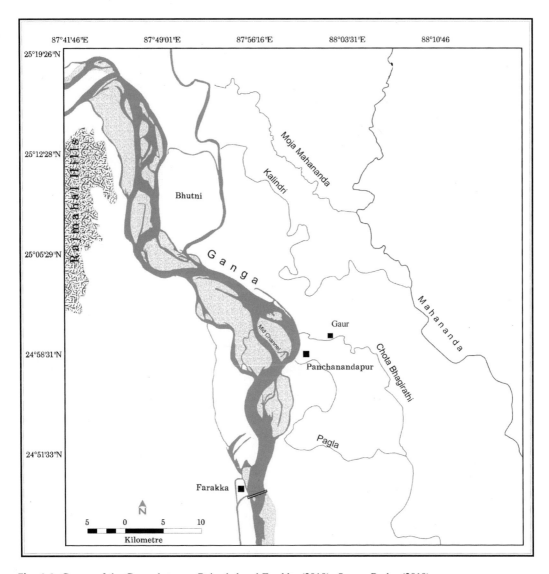

Fig. 4.4 Course of the Ganga between Rajmahal and Farakka (2010). *Source* Rudra (2010)

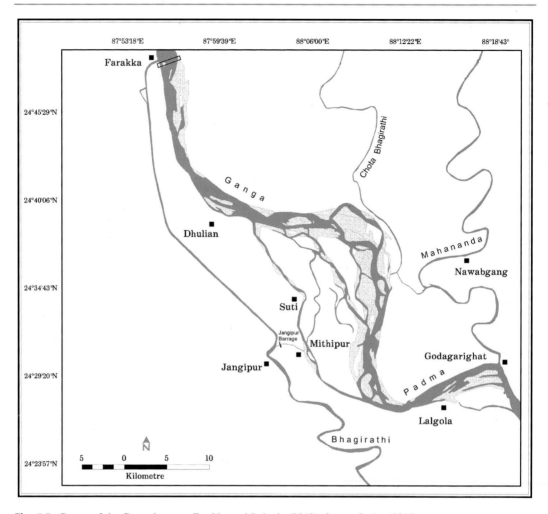

Fig. 4.5 Course of the Ganga between Farakka and Lalgola (2010). *Source* Rudra (2010)

unconsolidated sand of sand at the base of shelving cliff and effluent seepage of groundwater towards the river during post-monsoon season. The Ganga has been changing its course unabated from between Rajmahal and Jalangi, a length of about 185 km (Rudra 1996, 2000, 2010, 2014). The mighty river changed its course by avulsion prior to the late eighteenth century but now meander migration is the dominant process. Three moribund distributaries, namely the Kalindri, the *Chhota* Bhagirathi and the Pagla, mark the old course of the Ganga (Fig. 4.4). These channels were active when the Ganga used to flow along the mediaeval capital towns Gaur or Pandua. The Ganga subsequently deserted its old distributaries and now flows through one major and two other channels. A few islands locally known as *chars* have emerged between the channels. The Ganga has formed a mighty bend in Malda district of West Bengal and thousands of bank-dwellers lost their farmland and homestead. The river reached the ultimate limit of its eastward migration in 2007 and opened new route through the *char* and that is called mid-channel. In the Murshidabad district, the river was chosen as the Indo-Bangladesh border in 1947, but subsequently moved southwards (i.e. towards India) in some of the stretches (Fig. 4.5). But now the river

4.1 The Ganga through West Bengal

has again migrated towards Bangladesh. In fact, the Ganga/Padma migrates laterally along the international border, and its meander belt has a width of about 10 km (Fig. 4.14).

4.2 The Dynamic Ganga

A look at historical maps prepared and published during the period over last three centuries and comparing those with recent satellite images produced by the National Remote Sensing Agency (India) and the United State Geological Survey help to appreciate the dynamics of the Ganga in Bengal. But the maps published prior to Rennell (1780) were both cartographically and geographically unreliable. It is difficult to register those early maps on the GIS platform due to their wrong coordinates. D'Anville (1752), French cartographer, drew a map of Bengal and that looked better than what his predecessors produced; but that could not be treated as a correct map. James Rennell attempted first geographical survey of Bengal during 1764–1777, and the maps were produced in a compilation entitled *A Bengal Atlas* (1780). It was a collection of thirteen maps describing undivided Bengal. Colebrooke (1801), a successor of Rennell in Survey of India, prepared a map showing course of the Ganga through Bengal in 1796–97. Tassin (1841) made commendable contributions which help in the understanding the dynamic rivers of Bengal. The Survey of India was established in 1767. The Survey of India conducted revenue survey in the mid-nineteenth century and published some maps of Bengal describing geographical details and the then courses of the Ganga. More detailed survey was conducted in the early twentieth century, and geographical information was incorporated in topographical sheets, published in the scale 1:63,360. The topographical sheets were updated since 1970s and republished on a scale of 1:50,000. The National Atlas and Thematic Mapping Organization and the Survey of India carried forward the task of publishing district maps in 1990s. The National Remote Sensing Agency (India) provides false-colour satellite image which depicts latest geographical information. Bangladesh Space Research and Remote Sensing Organization also produced some land use and land cover maps. These maps when studied sequentially help to appreciate the changes in river courses of Bengal since the late eighteenth century.

4.3 The Changing Course of the Ganga

4.3.1 From Rajmahal to Farakka

While entering Bengal in the mid-eighteenth century, the Ganga had been flowing through two branches above Rajmahal and a mid-channel island named *Bhutni* stood in between. The principal branch flowed along the western side of the island. But when the Revenue Survey (1849) was conducted, the Ganga had changed in course and flowed along the north-eastern side of *Bhutni*. But the river subsequently changed the path and now flows along the south-western side of that island (Fig. 4.6). The older channel between the *Bhutni* and mainland has decayed. Further, Bhutni Island has taken a new look. The changes in the course of the Ganga between Rajmahal and Farakka can be visualized with reference to the ruins of Gaur which was the capital of Bengal for 155 years since 1203 AD, and its archaeological ruins along with the three moribund distributaries of the Ganga, namely Kalindri, Pagla and *Chhoto* Bhagirathi, are still traceable (Fig. 4.7) (Chapman and Rudra 2015).

Mukherjee (1938) in his 'The Changing Face of Bengal' noted '*Leaving the hills of Rajmahal the Ganges seemed to have passed northwards through modern Kalindri, and then southwards into the lower course of the Mahananda, east of the ruins of ancient Gaur*'. He borrowed information from the mediaeval maps of De Barros (1615) and Mukherjee (1938). Rennell (1793)

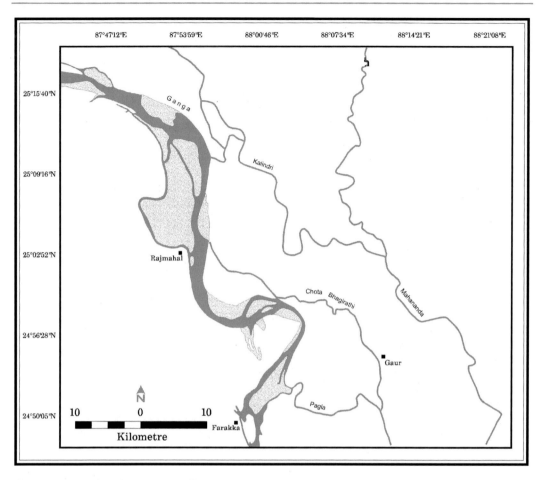

Fig. 4.6 Course of the Ganga between Rajmahal and Farakka (1849). *Source* Rudra (2010)

wrote in his Memoir of a Map of Hindoostan '*No part of the site of ancient Gour is nearer to the present bank of Ganges, than four miles and a half; and some parts of it, which were originally washed by the river, are now 12 miles from it. However, a small stream that communicates with the Ganges, now runs by its west side, and is navigable during rainy season. On the east side, and in some places within two miles, it is the Mahananda river; which is always navigable, and communicates also with the Ganges*'. Rennell's account confirms that the Ganga then flowed along the site of Gaur but migrated south-west subsequently. The old channel is now called *Chhoto* Bhagirathi. Creighton (1817) described the geography of the city of Gaur and quoted Manuel de Fariay Sousa's mid-sixteenth century description; '*Gaur, the principal city of Bengal, is seated on the banks of the Ganga, three leagues in length, containing one million and two hundred thousand families, and is well fortified. Along the streets, which are wide and straight, are rows of trees to shade the people, who are so very numerous, that sometimes many are trod to death*'. Creighton further noted '*The site of the city of Gaur is now an uninhabited waste on the eastern side of the Ganga, running nearly in a direction with it from S.S.E to N.N.W. The extent of the city appears, from the old embankments, (which enclosed it on every side) to have been ten miles long, and from one to one and a half broad. The banks (some of which are faced with bricks)*

4.3 The Changing Course of the Ganga

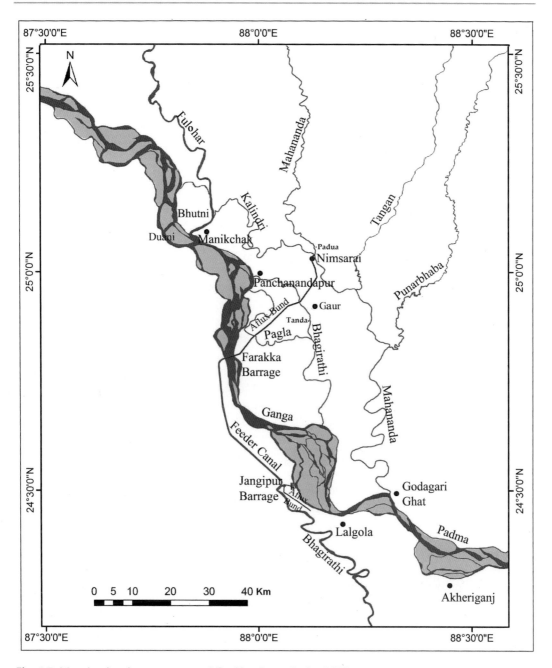

Fig. 4.7 Map showing river system around Farakka. *Source* Rudra (2010)

were sufficiently capable of guarding it from floods, during the height of the rivers, when the adjacent country was inundated.'

In 1358, the capital of Bengal was moved to Pandua which rendered better protection from the flood and erosion because of its elevated location on older alluvium of Barind tract. Khan and Stapleton (1930) explained *'The reason for the establishment of Pandua in the first instance seems to have been the existence of an island* of *barind (old red alluvium), close to the junction of the Mahananda river, and a former bed of the Ganga. Later, when the main course of the Ganga moved further south and the Mahananda, which still protects the town on the west, lengthened its course, Old Maldah at the junction of the Mahananda and the Kalindri (which took the place of the Ganga) became the actual port of Pandua.'* Chakravarti (1909) elaborated the issue of shifting the capital from Gaur to Pandua and noted *'The causes of this transfer are nowhere stated; but it was obviously connected with the changes in the river courses, making Lakhnauti (i.e. Gaur) unhealthy and uninhabitable. The various civil wars, with repeated plunderings of the city, might have hastened the transfer'*. The capital remained there for only 34 years and again returned to Gaur in 1392. The transfer was again governed by the changing course of the Ganga which again came back to the western side of the city. Stewart (1813) noted that, when Nasir Shah shifted the capital to Gaur, *'this Prince constructed the fortifications round the city of Gaur; the gates of which are still in existence; and the foundations have been traced over its whole circumference'* (Chapman and Rudra 2015). The decline of Gaur might be due to the tectonic causes. Hirst (1916) opined *'There was a severe earthquake in 1505 AD and shortly after it, the Ganges left its old course past Gour and retreated southwards.'*

But the Ganga this stretch was so dynamic that a favourable location of a town might be unfavourable after a single flood. So happened to the capital towns along the apex of the delta. The capital moved from Gaur to Tanda in 1564 and continued to stay there till 1591. Tanda was located at the confluence of the Pagla and the Chhoto Bhagirathi. The site was presumably engulfed by the gnawing river, and capital was shifted to Rajmahal. Ralph Fitch in 1585 (quoted in Foster 1921) noted *'Tanda standeth from the river Ganga a league, because in times past the river, flowing over the bankes, in time of raine did drowne the countery and many villages, and so they do remaine. And the old way the river Ganga was woont to run, remaineth dry, which is the occasion that the city doth stand so farre from the water.'*

Rajmahal, being located on the hard basaltic outcrop, offered a stable site for the capital town, but in 1608 Dhaka was chosen as the capital town for better connectivity and also for the strategic reason. The capital was again taken to Rajmahal in 1639 and brought back to Dhaka in 1660. It was taken to Murshidabad in 1702 and finally to Kolkata after the battle of Plassey in 1757. Kolkata continues to be the capital town of West Bengal and Dhaka is that of Bangladesh.

The navigational access to the capital town of Gaur, which flourished and declined between 1203 and 1564, was offered by the Kalindri, the *Chhoto* Bhagirathi and the Mahananda below Nimasarai. Satgaon and Gaur had trade link in the mediaeval period (Ray 2012; see Fig. 6.7 for location of Satgaon). The navigation route must have been through the Bhagirathi–Ganga–Mahananda–Kalindri or *Chhoto* Bhagirathi. Now one cannot sail up the present river route from Satgaon to Gaur. The archaeologists have identified both anchorage site and *snan* (bathing) ghat along the bank of *Chhoto* Bhagirathi which carried substantial waters of the Ganga and facilitated navigation. The lower Mahananda below Nimasarai was a part of the Ganga. The study of sediment collected from the banks of both Mahananda and Bhagirathi revealed that those are similar in nature with that carried by the Ganga. When the old Ganga flowed along the Gaur in the past, the present course of the Ganga between Rajmahal and Godagari Ghat of Rajshahi (Bangladesh) was possibly non-existent or it was an insignificant channel. When Rennell conducted the survey of Bengal during 1764–1777, the Ganga was flowing far away from Gaur.

In the second half of the eighteenth century, the Ganga flowed through a south-easterly course between Rajmahal and Farakka. Since the *Chhoto* Bhagirathi is still traceable along the western margin of the archaeological ruins of Gaur, it is understood that the Ganga did not sweep over the city of Gaur. It avulsed from one channel to other separated by a large 'interfluve' and possibly could have done many times. The revenue survey map (1847–49) showed a robust meander of the Ganga in Malda (Fig. 4.6) but the river again adopted a straight path during the early twentieth century. The construction of a barrage across the Ganga at Farakka started in 1962–63 and was completed in 1971. Consequently 87 million m^3 of water was stored in the barrage-pond, and the water level swelled about 6.71 m. The river widened itself to accommodate impounded water, and impact was observed even beyond Bhagalpur (Parua 1999, 2002). The Ganga again started to migrate eastwards and a mighty bend was formed. That bend was shown in the topographical sheet (1970–71) of the Survey of India and that of US Air Survey 1982. The bank erosion by the Ganga took alarming magnitude and continued till rainy season of 2006, and thousands of villagers were displaced. In course of eastward migration, the Ganga has so far eroded about 267 km^2 fertile land in Malda. As a corollary, newly deposited land, locally called *char*, has surfaced along the opposite bank. Since 2007, the main flow of the Ganga has been passing through a newly opened channel through the *char*, and that channel is called as the 'mid-channel'. Notably while Rajmahal is geologically resistant to erosion and Farakka is also a nodal point where the river has not moved.

This motivated engineers to choose Farakka as the site of barrage (Figs. 4.8 and 4.9). It is appeared from the recent satellite image (Fig. 4.9) that the Ganga has been discharging through four outlets below Rajmahal, and the mid-channel carries bulk of the discharge now. The flowing water impinges the left bank just upstream of the barrage and erodes land. The engineers are apprehensive the Ganga may outflank the Farakka barrage and open a new route towards Bangladesh with disastrous consequences of much greater magnitude than what was caused by the Kosi in 2008.

4.4 Changing Course Between Farakka and Jalangi

The lateral oscillation of the Ganga downstream of the Farakka continues unabated, and there are many published documents and maps which help to appreciate this change (Rudra 2010, 2014). D'Anville (1759) in his 'Geographical Illustration' described Rajmahal as the most important town of Bengal. He noted '*Its situation on the Ganges is very remarkable, being at the place where the river divides into two principal branches, through it runs into the sea, about 70 leagues lower, forming the delta more considerable than that of Nile, and of which Raji-mohol is the top…. Of these two branches of river, one is called the Great, and the other Little Ganges: The great one is that on the left hand going down it, and leads to Daka: It is however less known than that on the right hand, on which the Europeans have erected settlements; and which is their usual channel to go up into the country.*' The author described the Ganga–Padma flowing through Bangladesh as 'Great' and the Bhagirathi–Hugli river flowing southwards through West Bengal as 'little Ganges'. This branch was joined by another spill channel of the Ganga taking off at Suti (see Fig. 6.2). The evolution of the off-take of the Bhagirathi is discussed further in this chapter.

The Revenue Survey of Murshidabad was conducted in the mid-nineteenth century, and the Ganga between Farakka and Suti had been flowing through a great bend during that time (see Fig. 4.12). But the Ganga altered its course and moved south-west during subsequent decades. Notably the course of *Chhoto* Bhagirathi now flows along the former course of the Ganga (see Fig. 4.7). Since the construction of the Farakka barrage in 1975 and interception of the flow, the hydraulic gradient of the Ganga has been altered and relatively silt-free water flowing downstream incised the bed deeply, and the channel has become progressively deeper and

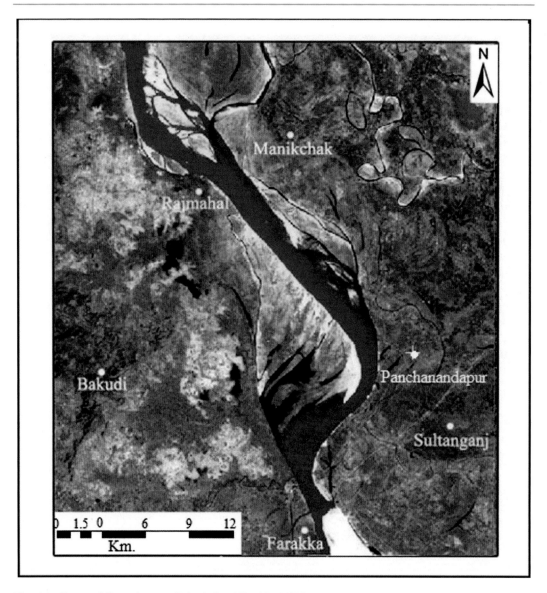

Fig. 4.8 Course of Ganga between Rajmahal and Farakka (1972)

narrow. The water released from the barrage impinged the right bank which gradually migrated further west. This change is clearly discernible if one compares the revenue survey map (1849), US air survey map (1982) and the recent satellite images (2010). Even base of the Farakka Barrage is deeply scoured, and the engineers dump boulders just downstream of the barrage to protect the same from being dislodged.

In the decade of 1960 and early 1970s, the right bank of the river just down the Farakka barrage was severely eroded and about 20 km stretch between Dhulian and Suti experienced rapid erosion. The railway and national highway connecting North and South Bengal were delinked and subsequently realigned further away from the river bank. In fact, present Dhulian town, an important trading centre of the

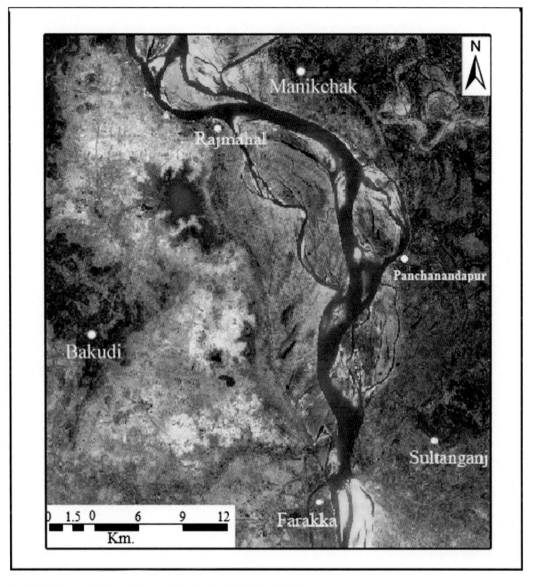

Fig. 4.9 Course of Ganga between Rajmahal and Farakka (2010)

district, is the new site as the former was wiped out. A narrow interfluve separates the Ganga and the Bhagirathi and the gnawing Ganga came close the Bhagirathi in 1990s. The width of the interfluve along a straight line was 2.86 km in 1976 but was reduced 1.40 km in 1996. It was further reduced to 800 m in 2000 (Rudra 2004). This deepened the anxiety of the Government because unification of two rivers might make the Farakka barrage Project redundant. But fortunately for India, the Ganga again went towards opposite bank during subsequent years (see Fig. 4.15).

Akheriganj, located further east, was in peril in the late 1980s and early 1990s (Figs. 4.10 and 4.11). The impinging river engulfed the residential and commercial areas. The only road connecting the area with the district headquarters

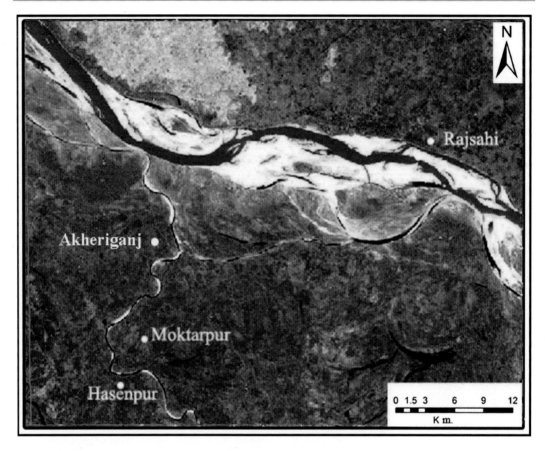

Fig. 4.10 Course of Ganga near Akheriganj in Murshidabad district (1972)

was delinked. The bank revetment works failed to protect the villages from erosion. The present cluster of settlements, also known as Akheriganj, has developed further inland, and the older site stands on the opposite bank of the river is at least 3 km north of the new site. Many neo-refugees have migrated to *Nirmal Char* which newly resurfaced along the bordering Bangladesh. More than 200 km^2 of land in Murshidabad was eroded during the period 1988–1994 causing displacement of 79,000 people, and land covering equal area has emerged along opposite bank (Rudra 1996). In 2002, the Ganga went back to its older course and started to flow through the old channel which delineates the Indo-Bangladesh border near Akheriganj (Fig. 4.12).

In 1994, Jalangi town, located further east, was severely eroded, and the local market, *panchayat* (village Government) office, high school building, police station and 450 houses were swallowed by the Ganga. Even the border road connecting Karimpur was disconnected. The main flow of Ganga after reaching the limit of its southward encroachment shifted northwards and left behind newly emerged land (Figs. 4.13 and 4.14).

4.5 Mechanism of Changing Course

The Ganga drains a catchment covering more than one million km^2 and continuously transfers water and sediment load into the sea. The Ganga

4.5 Mechanism of Changing Course

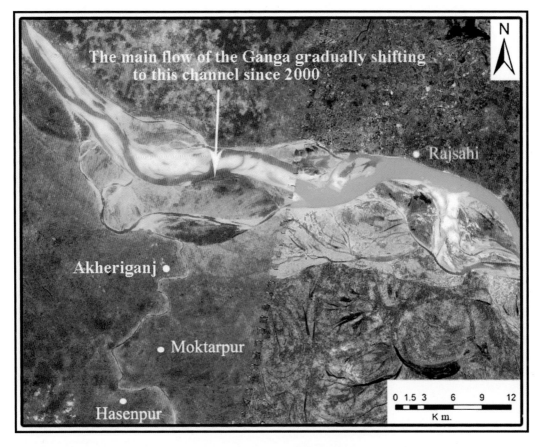

Fig. 4.11 Course of Ganga near Akheriganj in Murshidabad district (2010)

is unique river in terms of its fluvial geomorphology. It is revealed from estimate of flow that approximately 80% of annual volume of water passes during four monsoon months (June to September). In September 1998, large part of Malda district was submerged and historically highest discharge of about 2.70 million cusec was recorded at Farakka. In years of normal rainfall, peak discharge at Farakka does exceed two million cusec. But lean season's discharge may be less than 55,000 cusec exerting stress over the Indo-Bangladesh hydro-diplomacy (Rao 1979). Many structures built across the Ganga in upper riparian Indian states and Nepal to ensure lean season irrigation leads to downstream reduction of flow and one can even walk across the Ganga near Allahabad in the month of February. The Kosi, the Gandak, the Ghaghra and other rivers draining from the Nepal discharge into the Ganga and rejuvenate it again (Hollick 2007).

The experts have put forward different estimates about sediment load in suspension carried by the Ganga per year. It was estimated as 800 million tons by Abbas and Subramanium (1984), 794 million ton by Wasson (2003) and 736 million ton by Rudra (2006). Tadan et al. (2008) ranked the Ganga second position worldwide in terms of total suspended load (524MT/yr). Wasson (2003) noted that about 90% of the sediment load carried by the Ganga is contributed by the left bank tributaries draining the Himalaya and less than 10% by right bank tributaries coming from peninsular plateau. The volume of water and

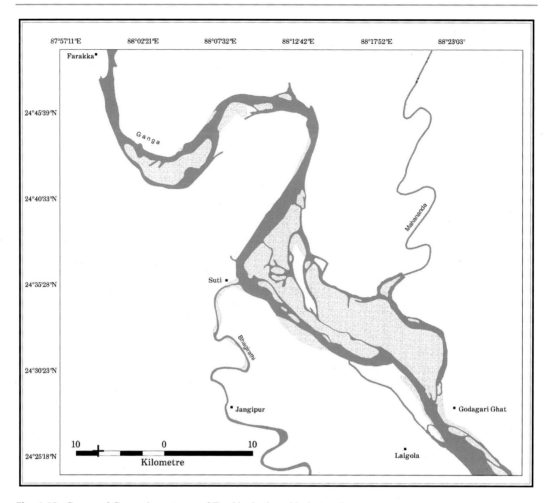

Fig. 4.12 Course of Ganga downstream of Farakka in the mid-nineteenth century

sediment load are important components of fluvial dynamics especially the meander geometry. The human interventions into the fluvial regime in the forms dams/barrages or bank protection work render impacts on the functioning of the river and the processes of erosion, transportation and deposition may be delayed or accelerated. In an uninterrupted situation, the dynamics of the river is governed by volume of flowing water, bank stratigraphy, scouring of bank and effluent seepage of water. The wavelength and amplitude of a meandering are related to the highest discharge flowing through the river which becomes wider during the monsoon months and shrinks again in lean months.

The shelving bank of Ganga is composed with the layers unconsolidated sand with intervening layers of silt-clay. The fast current of the Ganga during the monsoon removes sand from the base of bank leading to its collapse. The groundwater is recharged through the porous bank when the river is bank-full. The post-monsoon effluent seepage towards the river creates void under the shelving bank and leads to collapse (Rudra 2000). The attempt to protect bank with boulders is an expensive technology but often failed to ensure security against the problem of erosion. The piecemeal bank protection causes accelerated erosion on both upstream and downstream of the protected stretch.

4.5 Mechanism of Changing Course

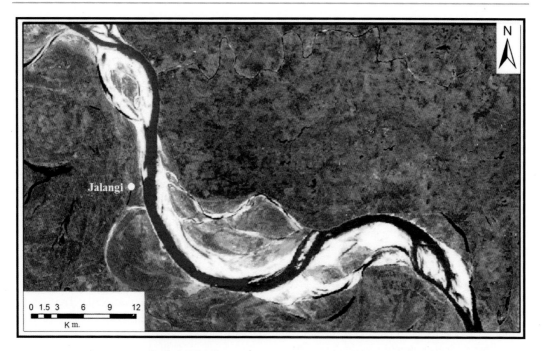

Fig. 4.13 Course of Ganga near Jalangi (1972)

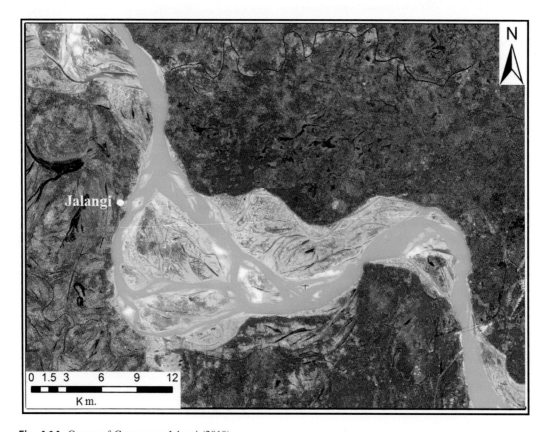

Fig. 4.14 Course of Ganga near Jalangi (2010)

Table 4.1 Estimated flow of the Ganga/Padma at Jalangi (MCM)

River	Jan	Feb	Mar	Apr	May	June	July	Aug	Sep	Oct	Nov	Dec
Ganga–Padma (at Jalangi)	12,147	10,944	8369	9044	13,188	104,259	109,186	108,485	104,599	12,621	10,963	11,289

4.6 Estimated Flow

The discharge flowing through the Ganga reaches the highest level in the August, and minimum flow is recorded in March (Table 4.1). Since data relating to discharge in all transboundary rivers are treated as 'classified' by the Central Water Commission, Government of India and the researchers are not allowed to access recorded data, the only option is to appreciate the mean monthly quantum of water in the Ganga through a mathematical model. The model estimates the discharge draining out of the basin. The Table 4.1 furnishes the estimated monthly volume of water flowing through the Ganga at Jalangi (the last town on the Indo-Bangladesh border). The data generated through the model do not necessarily mean real-time data but found to be close to it on field verification.

4.7 The Ganga System in Bangladesh

The Ganga in Bangladesh changes its course so frequently that it is described as *kirtinasha* or destroyer of creation. It is observed from the revenue survey map of the mid-nineteenth century that the Ganga threw off a distributary called Boral (at Charghat near Rajshahi) which flowed eastwards to the Jamuna after joining with the Karatoya at Bera. This channel is now in moribund condition and disconnected from the Ganga. The tract lying between Indo-Bangladesh border in the west and the Ganga–Padma in the north, the lower Meghna in the east and Bay of Bengal in the south is drained by a complex network of distributaries. Of these, the Gorai-Madhumati and the Arialkhan are most important. In addition, this tract is drained by many beheaded channels. These are the Morichap, the Kobadak, the Bhairab, the Chitra, the Nabaganga, the Kumar, the Sitalakha, the Gahgar, Bishakandia and many other minor streams. All these rivers flow in a south-eastward direction. The estuaries through which waters ultimately reach the sea are the Jamuna, the Malancha, the Pusur, the Baleswar, Patuakhali and the mighty Meghna. The Bangladesh Water Development Board (Govt. of Bangladesh 2005) has listed 98 rivers draining this region, and each river was assigned an identification number (Fig. 4.15).

All the distributaries of the Ganga are decaying fast. The decay of the Bhagirathi, the Jalangi and the Mathabhanga–Churni–Ichhamati are generally ascribed to easterly flight of bulk of discharge of the Ganga (Morgan and McIntire 1959). After the decay of the Bhagirathi since the early seventeenth century, the Jalangi and Mathabhanga continued to be navigable. Rennell (1793) found Jalangi navigable for about nine months of the year, while in 1797 Colebrooke (1801) described the Mathabhanga more easily navigable than the Bhagirathi and the Jalangi. During the first half of the nineteenth century, the Garai became rejuvenated at the expense of the Jalangi and the Mathabhanga. The Garai which had a width of about 183 m in 1828 was widened to 582 m in 1856 (Sherwill 1858). The lower course of the Garai which was first known as Alankhali and then as Madhumati further downstream was so narrow that 'it could be jumped over by a man on a horseback', but in 1830–1833 it was described as finest deltaic outlet opened round the year for navigation (Fergussion 1863).

4.7 The Ganga System in Bangladesh

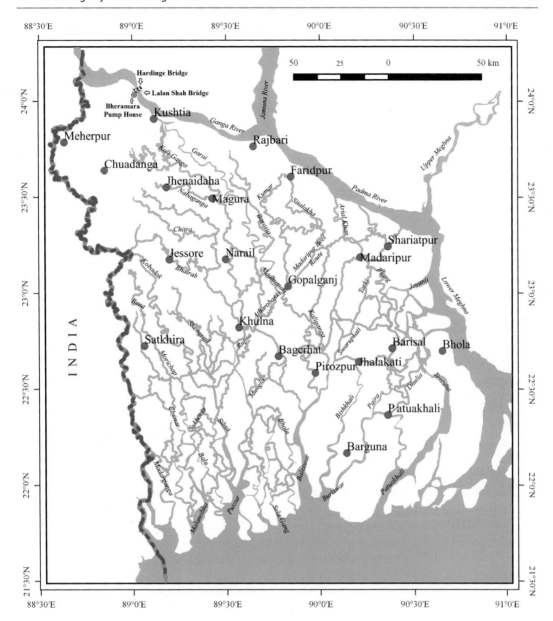

Fig. 4.15 Distributaries of the Ganga in Bangladesh. *Source* BDPUB (2011)

Being reinforced by the waters of the Navaganga, the Madhumati River (lower Garai) widened to nearly 5 km and became the second largest deltaic outlet after the Meghna estuary. Both Sherwill (1858) and Fergussion (1863) presumed the possibility of rejuvenating the Garai and silting up of the fordable portion of the Ganga–Padma lying east of Kusthia where the Garai takes off. But their presumptions subsequently proved futile, and the Garai became moribund during the early twentieth century (Ascoli 1910). The rejuvenation of the Garai was attributed to the effect of hydraulic damming caused by the floodwater of the Brahmaputra

(Jamuna) which reaches about a month earlier at Goalundo than the Ganga. It was then estimated that if water level at Goalundo gains an additional height of 2 m, the impact of back flow through the Ganga would reach up to the Garai off-take at Kusthia, about 70 km upstream. Had the water-dam been so effective in opening the intermediate distributaries, it would have opened all the distributaries successively from the east to west. But such thing never happened, and the Chandana, a distributary lying east of the Garai, was found dry by Fergussion in 1863. Till the first half of the nineteenth century, the Ganga/Padma and the Jamuna had separate outlets into the Bay of Bengal (Fig. 5.1). Since the Jamuna altered its course in 1830 and flowed southwards at the expense of old channel through Sylhet basin, the two rivers were united at Goalundo. The combined flow joined with the Meghna at Chandpur and discharged into the sea through the Shahbazpur estuary.

4.8 Policy Issues

The Ganga flowed along interstate boundary between Bihar (now Jharkhand) and Indo-Pakistan (now Bangladesh) border in 1947. The Survey of India noted in 1946 that *'the Province and district boundaries in the Ganga river follows the main deep water channel and will vary as the course of deep water channel changes.'* The British rule ended in 1947, and it was declared *'the territorial extent of the State of West Bengal and also of Bihar stands fixed since the commencement of the Constitution and cannot be altered because of any change in a river running along interstate boundary, except by way of amendment of the Constitution'* (vide DL. R & S. W. Bengal Memo Nos. 145/383/D/Con/90 dt. 31.7.1992; 145/493/D/90 dt. 17.12.1999 & 145/360/D//90 dt 9.9.2003).

In course of arbitration on Indo-Pakistan boundary commission, Bagge (1950) left no ambiguity and decided that *mid-stream of the river Ganges from the point little below Farakka to the point where Mathabhanga takes off from the river Ganges was taken to be the boundary between Murshidabad and Rajshahi. It was declared that irrespective of the changes of the course of river the boundary should remain fixed.* After the independence in 1947, fixing the interstate and international river border has posed problems due to lateral oscillation of the Ganga.

The Ganga migrated eastwards away from Jharkhand–West Bengal border, and more than 267 km^2 of land belonging to Malda district (West Bengal) have been eroded and subsequently resurfaced along the mainland of Jharkhand (Rudra 2012). This land is attached with the mainland of Jharkhand. More than 0.10 million erosion victims migrated to the newly emerged land along opposite bank. Since natural boundary wiped out and no survey was conducted till date, possession of land has been an issue of conflict among settlers. The Government of West Bengal does not recognize the new clusters of settlements as revenue villages. The settlers are denied minimum civic amenities.

The southward migration of the Ganga leaving the Indo-Bangladesh border has displaced many people who have migrated further inland or to newly emerged land along the mainland of Bangladesh. A group of neo-refugees has settled on the mainland further south. The two Experts' Committees (P. Singh Committee in 1980 and G. Keskar Committee in 1996) constituted to recommend the remedial measures recommended for revetment of bank with boulders, but the issue of rehabilitation of neo-refugees was grossly denied.

The river management in both India and Bangladesh has been guided by a narrow sectoral outlook largely based on civil engineering approach of bank protection. But it is more important to prepare a scientific land use/land cover plan for the meander belt of the Ganga which should be declared as the 'space for the river'. It was suggested by Experts' Committee (1980) that bank protection work is so expensive that measures can only be taken to protect national assets like railway, highway and important urban centres. The best survival

4.8 Policy Issues

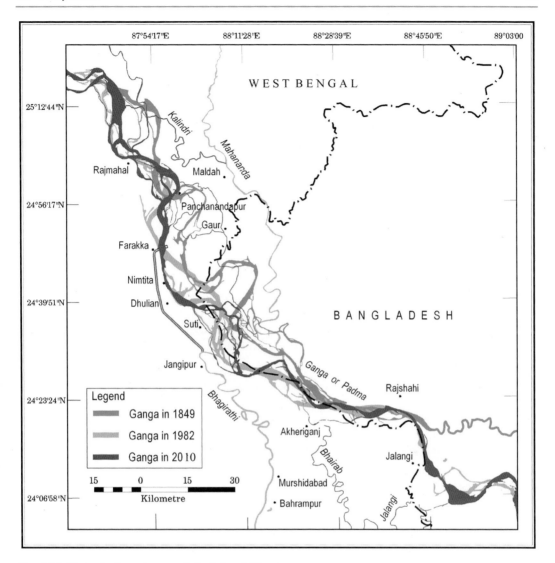

Fig. 4.16 Dynamic Ganga between Rajmahal and Jalangi

mechanism to live with flood and erosion is allowing the river to migrate both ways within its meander belt and planning the land use and land cover accordingly (Figs. 4.16 and 4.17). The river bank can best be used as agricultural land and not for any permanent structures or alignment of road/railway. Notably river banks are most densely populated parts of this subcontinent, and such planning will require rehabilitation of many people.

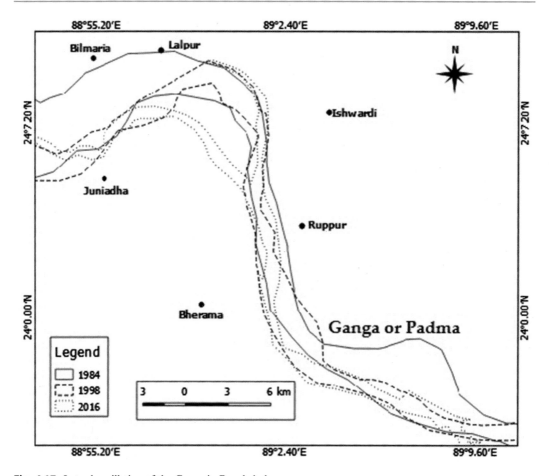

Fig. 4.17 Lateral oscillation of the Ganga in Bangladesh

References

Abbas N, Subramaniam V (1984) Erosion and sediment transport in the Ganga River Basin (India). J Hydrol 69:173–182

Ascoli (1910) Rivers of the Delta, J Asiatic Soc Bengal 543–556

Bagge A (1950) Report of the international Arbitral Awards. Boundary disputes between India and Pakistan relating to the interpretation of report of Bengal Boundary Commission, Part I, published by UNO. (www.pib.nic.in)

Bangladesh Pani Unnyan (Water Development) Board (2011) Rivers of Bangladesh (in Bengali)

Chapman GP, Rudra K (2015) Time streams/ history and rivers in Bengal. Centre for Archaeological Studies & Training, Eastern India

Chakravarti M (1909) Notes on Gaur, etc. J Proc Asiatic Soc Bengal vol V, 1909 no. 7, pp 204–234). (Khan 1930:17)

Colebrooke RH (1801) On the Courses of Ganges through Bengal. Asiatic Res 7:1–31. (Calcutta)

Creighton H (1817) The Ruins of Gour, London

D'Anville JBB (1759) A geographical illustration of the Map of India; Printed for the Editor at the Golden-Globe, London, p 29

de Barros J (1615) Description of the Kingdom of Bengal

Fergussen J (1863) Recent changes in the delta of the Ganges. Q J Geological Soc (of London) 19:321–354

Government of Bangladesh (2005) Bangladesh Water Development Board (2005) Bangladesher Nad-Nadi, Dhaka, pp 425–620

Foster W (1921) Early travels in India 1553–1619. Humphrey Milford/Oxford University Press, p 24

Hirst FC (1916) Report on the Nadia Rivers. The Bengal Secretariat Book Depot, Calcutta

Hollick JC (2007) Ganga. Random House, Delhi, pp 13–98

Keshkar G et al (1996) Report of experts' committee for Bank Erosion Problem of River Ganga-Padma in the districts of Malda and Murshidabad, Planning Commission, Govt. of India, 1–71

Khan M, Abid A, Stapleton HE (1930) History and Archaeology of Bengal: or Memoirs of Gaur and Pandua. Calcutta

References

Mishra DK (2008) Finger in the dike. Himal South Asian 21(12):44–46

Mukherjee RK (1938) The changing face of Bengal/a study in riverine economy. University of Calcutta, p 97

Morgan JP, McIntire WG (1959) Quaternary Geology of the Bengal Basin, East Pakistan and India. Bull Geolog Soc Am 70:319–342

Parua PK (1999) Erosion problem of the River Ganga in the districts of Maldah and Murshidabad in West Bengal. Civil Engineering Today, ASCE, Calcutta X111(2):3–20

Parua PK (2002) Fluvial Geomorphology of the River Ganga around Farakka. J Inst Eng 82:193–196

Rao KL (1979) India's Water Wealth. Orient Lonman, Kolkata, p 213

Ray A (2012) The City of Gaur; In Gaur/The Mediaeval City of Bengal; (Pratna Samiksha, A J Archaeol, Special Issue), 83–113

Rennell J (1780) A Bengal Atlas. In: Rudra K (ed) (2016) and published by Sahitya Samsad, Kolkata

Rennell J (1793) Memoir of a Map of Hindoostan (Reprinted in India in 1976), p 147,385

Rudra K (1996) Problems of Bank Erosion in Murshidabad District of West Bengal. J Geography Environ vol 1. Vidyasagar University, West Bengal, pp 25–32

Rudra Ka (2000) Living on the edge: the experience along the Bank of the Ganga in Malda District, West Bengal, vol 5. Indian J Geography Environ, Vidyasagar University, West Bengal, pp 57–66

Rudra K (2004) Ganga-Bhangan Katha (in Bengali) The Saga of Ganga Erosion, Kolkata, pp 85–95

Rudra K (2006) Shifting of the Ganga and Land Erosion in West Bengal/A Socio-ecological Viewpoint. CDEP Occasional Paper-8. Indian Institue of Management, Calcutta, pp 1–43

Rudra K (2010) Dynamics of the Ganga in West Bengal, India (1764–2007): implications for science–policy interaction. Q Int 227(2):161–169

Rudra K (2009) Re-flooding the Kosi. Himal South Asian 22(3):50–51

Rudra K (2012) Atlas of the Changing River Courses in West Bengal. Sea Explorers Institute, Kolkata

Rudra K (2014) Changing River Courses in the Western Part of the Ganga-Brahmaputra Delta Geomorpholgy 227:87–100

Sherwill (1858) Selections from the Records of the Bengal Government/Reports on the Rivers of Bengal; G.A. Savielle, Calcutta Printing and Publishing Company

Singh P et al (1980) Report of the Ganga Erosion Committee, Govt. of West Bengal, 1–43

Sinha R (2008) Kosi: rising water, dynamic channel and human disaster. Econ Political Week 15:42–46

Stewart C (1813, references above to 1903 reprint) The History of Bengal: from the first Muhammeden Invasion until the Virtual Conquest of that Country by the English A.D. 1757. Calcutta

Tandon SK, Sinha R, Giblin MR, Dasgupta MR, Ghanzanfari AS (2008) Late quaternary evolution of the ganga plains: myths and misconceptions, Recent Development and Future Directions. Golden Jubilee Memoir of GSI, No. 66, pp 259–299

Tassin JB (1841) The New Bengal Atlas. Calcutta

Wasson RJ (2003) Sediment budget in the Ganga Brahmaputra Catchment. Curr Sci 84(8):1041–1047

The Jamuna–Meghna System

5

Abstract

The Jamuna–Meghna system drains the Barind tract and Sylhet basin of Bangladesh. The changes in river courses within this subsystem since the late eighteenth century are most striking. Earlier the Ganga and the Jamuna discharged separately into the Bay of Bengal. The Teesta avulsed eastwards in 1787. The Brahmaputra left its course through Sylhet basin 1830 and adopted a then insignificant channel called Janai. Since then the lower Brahmaputra is known as Jamuna. The Meghna extended itself to join the Ganga or Padma at Chandpur. The subsidence of central Bengal (i.e. along the present course followed by the Jamuna) acted as the cardinal force governing the changes. The absorption of sediment load carried by the old Brahmaputra in the Sylhet basin for a longer geological period may be attributed to the disproportionate growth of the delta towards the Bay of Bengal.

The Jamuna divides the Bengal basin into two unequal halves. The course presently followed by the Jamuna in Bangladesh is supposed to be a subsiding trough leading to the convergence of many rivers (Morgan and McIntire 1959). The Meghna system drains the area lying east of the Jamuna and the Ganga and some tributaries to the former (Jamuna) drain the western region. The Ganga and the Meghna are primarily meandering in nature but the Jamuna is a braided river with numerous interlacing channels separated by point bars lying in between. It travels about 230 km between Dhubri and Goalundo where it meets the Ganga. The width of Jamuna varies between 3 and 10 km. The maximum discharge flowing through *this river during the monsoon* may be 71,000 m^3/s. The comparatively steeper slope and huge bed load compel the river to be braided. It is also stated by some experts that the devastating earthquake of 1950 in Assam generated a huge sediment load which was transferred and deposited in the downstream stretch (Sarkar et al. 2013). The braided channels of the Jamuna and shapes of numerous point bars change continuously during the monsoon when 13 million tons of suspended loads is transported downstream/day (Coleman 1969).

The Ganga/Padma and Jamuna/Brahmaputra discharged independently into the Bay of Bengal till the early nineteenth century (see map of Rennell 1780). Notably, they were outlets of two separate basins (Fig. 5.1). The Jamuna flowed through the Sylhet basin to receive the Meghna as the left bank tributary and ultimately opened the way to the sea keeping Shahbazpur (now known as Bhola) island on the west; the Ganga had been reaching the sea through an estuary presently called Arial Khan–Tentulia keeping the said island on the east. The Brahmaputra–Meghna system played a significant role in depositional history of the Bengal basin, especially in the filling up of the Sylhet trough. Goodbred, Jr. and Kuehl (2000) explored the changing courses of the Brahmaputra which flowed alternately along both sides of Madhupur tract during the late Holocene period. They traced the history of avulsive and migratory channel behaviour during preceding 11,000 years from the stratigraphic records. The Brahmaputra flowed along western fringe of Madhupur tract during 11,000–7000 years BP and adopted an eastward course through the Sylhet basin during the mid-Holocene (7000–4000 years BP). It again migrated westward and adopted the old course during the late Holocene (<4000 years BP). But the channel through Sylhet basin remained active as well. Subsequently, it had been rejuvenated at the expense of the western channel but again went back to adopt its western course in 1830. This westward avulsion of Jamuna brought about some interconnected changes. The interfluve between the Jamuna, the Padma and the Meghna has a complex drainage network of sixty rivers. There are five distributaries originating from the left bank of the Jamuna and ultimately discharging into the Meghna through a common outlet. These rivers are the Jhinai, the Pungli, the Lowhaganj, the Ichhamati and the Dhaleshwari. The distributaries taking off from the old Brahmaputra are the Bangshi, the Banar, the Aiman-Akhila, the Panagariya-Shila and the Shitalakhya. These rivers also have a combined outfall into the Meghna near Munshiganj. In the early nineteenth century, the Dhaleshwari took off from the Ganga at Jaffargang (now known as Daulatpur) and flowed eastward to meet Meghna at Munshiganj. A branch of the Dhaleshwari called Buriganga flowed along Dhaka and re-joined the parent river (Fig. 5.1). Since the Jamuna altered its course in 1830 and joined the Ganga at Goalundo, the old site of Jaffargunj was eroded but the Dhaleshwari continued to flow as a branch of Jamuna taking off at Kalihati (Tangail District). This change happened between 1870 and 1920 (Coleman 1969). The Bansi having a beheaded off-take from the old Brahmaputra flows south-east to join the Dhaleshwari south of Savar. The river flows through Tangail Sadar, Basail, Kaliakoir, Tongi, Dhamnai and Savar. It appears from the satellite images that the Bansi in the remote past took off from the old Brahmaputra near Jamalpur and flowed along the western front of the Madhupur tract. The Shitalakhya also originates from the old Brahmaputra near Monohordi and flows about 108 km through Kapasia, Kaliganj, Rupganj, Narayanganj and Sonargaon before it discharges into the Dhaleshwari near Bandar (BDPUB 2011) (Fig. 5.2).

The meeting of the Padma and the Meghna happened when the minor cross-channel connecting two estuaries near Rajganagur was gradually enlarged and a common outlet was formed south of Chandpur (see map of Rennell). The process of unification took several decades of early nineteenth century and was completed before 1841 when J. B. Tassin published a map of Bengal where the common estuary of the Padma, the Jamuna and the Meghna was shown. The earlier course of the Ganga below Rajanagar was reduced substantially but continues to flow till date as Arial Khan (Sarkar et al. 2011).

The Meghna along with its tributaries drains the eastern part of Bengal basin. Its basin area covers 82,000 km^2 where the rainfall is reportedly the highest in this subcontinent and consequently the river carries 150 BCM of water annually. The peak discharge during the monsoon may be close to 12,000 cumec when the river carries 0.20 million tons of suspended solid (Coleman 1969). The Barak, the main headstream of the Meghna, originates from the northern hills of Manipur and flows for about 560 km in a sinuous course till it reaches Bhanga Bazar of Karimpur District where it is bifurcated into two branches—Surma in the north and Kusiyara in the south. Both flow westward and

5 The Jamuna–Meghna System

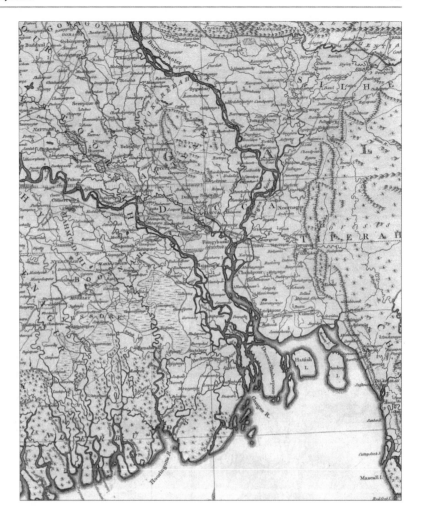

Fig. 5.1 Rennell's map (1780) showing the lower courses of the Ganga, the Jamuna and the Meghna

rejoin at Kuliarchar (in Kishorganj) to form Meghna. The Surma receives many tributaries draining the southern slope of Meghalaya Plateau. The rivers which join Surma as right bank tributaries are Piyan, Dhala, Nayagang, Jalukhali, Jadukata, Someswari, Kangsa and Chillakhall. Sari-Gowain. The old channel of Brahmaputra also joins Meghna at Bhairab Bazar.

The Kusiyara, the other branch of Barak, flows west and comprises a sub-basin with tributaries like Juri, Manu and Lungla draining from the Tripura Hills. The combined flow of Kusiyara and Surma travels southward as the Meghna and receives Khowai, Sutang, Sonai, Haora, Salda, Gumti and Kakri–Dakatia as left bank tributaries. It joins the Padma at Chandpur and flows into the Bay of Bengal through a mighty estuary between Bhola and Lakshmipur. The Feni, Muhuri and Selonia drain slopes of Khagrachhari and separately discharge into the Bay of Bengal. The Karnafuli, the most important river of Chittagong Hill discharges into the sea further south.

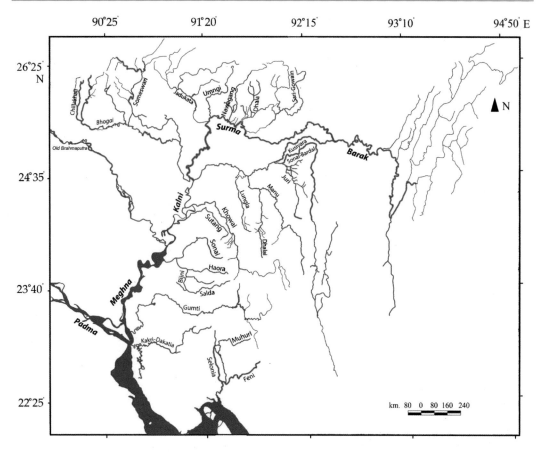

Fig. 5.2 The Meghna system in Bangladesh

References

Bangladesh Pani Unnyan (Water Development) Board (2011) Rivers of Bangladesh/North Central Region. (A Report in Bengali)

Coleman JM (1969) The Brahmaputra river, channel processes and sedimentation. Sed Geol 3(2, 3):123–239

Goodbred SL Jr, Kuehl SA (2000) The significance of sediment supply, active tectonism, and eustasy on margin sequence development: late quaternary stratigraphy and evolution of the Ganges-Brahmaputra Delta. Sed Geol 133:227–248

Morgan JP, McIntire WG (1959) Quaternary geology of the Bengal Basin, East Pakistan and India. Bull Geol Soc Am 70(3):319–342

Rennell J (1780) A Bengal atlas. Edited and Compiled: Rudra K (2016). Sahitya Samsad, Kolkata

Sarkar MH, Aktar J, Ferdous MR, Noor F (2011) Sediment dispersal processes and management in coping with climate change in the Meghna estuary, Bangladesh. Sediment problems and sediment management in Asian River basins. In: Proceedings of the ICCE workshop held at Hyderabad, India, 2009. IAHS Publication, 349

Sarkar MH, Aktar J, Rahman MM (2013) Century-scale dynamics of the Bengal Delta and future development. In: Fourth International Conference on Water and Flood Management, pp 91–104

The Bhagirathi-Hugli River System

Abstract

The Bhagirathi-Hugli River is the western branch of the Ganga and flows more than 500 km through West Bengal. The Jalangi and the Mathabhanga–Churni are two other offshoots of the Ganga, and those two join the Bhagirathi and Mayapur and Payradanga, respectively. The lower 280 km stretch of the Bhagirathi is tidal and known as the Hugli River. The Bhagirathi remains delinked from the Ganga for about nine months of the year and receives water from 38-km-long feeder canal originating from the Farakka barrage. The river tends to oscillate laterally in its non-tidal regime and has thrown several distributaries in the lower reach. The channels which had been important navigational route during the mediaeval and post-mediaeval period have gone dry. The attempt to resuscitate the Bhagirathi-Hugli River by inducing water through a feeder canal originating from the Farakka barrage has not worked to the level of expectation, and the river is in alarming state of decay.

The Ganga is divided into two major channels, namely the Bhagirathi and the Ganga/Padma about 40 km downstream of Farakka at a village called Mithipur (near Jangipur) of Murshidabad district, (Fig. 4.2). The Bhagirathi has a length of 500 km extended through West Bengal and ultimately disappears in the Bay of Bengal through a funnel-shaped estuary (Fig. 6.1). The tides penetrate through this river and reach Nabadwip, located 280 km north of the sea. The non-tidal regime, lying further north, has a length of 220 km. The Bhagirathi remained delinked from the Ganga except during the months of July to September and was virtually fordable in its upper reach. It was artificially rejuvenated in 1975 when a feeder canal connected the Ganga with the Bhagirathi and induced water from barrage pond (Rudra 2012).

The Bhagirathi has an undulating catchment area of 66,000 km^2 along right bank drained by eight major tributaries which are Bansloi, Pagla, Mayurakshi, Ajay, Damodar, Rupnarayan, Khori and Haldi. These tributaries together contribute about 48,410 million m^3 of water annually into the Bhagirathi-Hugli River. The eastern part of the catchment covering 5971 km^2 is drained by two other distributaries of the Ganga, namely Bhairab–Jalangi and the Mathabhanga–Churni (Fig. 4.2) which contribute 2922 million m^3 of water annually (Rudra 2014). These two channels do not get freshwater supply from the Ganga barring two or three months. There are two major sources of freshwater draining into the Bhagirathi-Hugli. The monsoon rain supplies substantial water from the catchment, and the groundwater contribution during lean months is also important. The lower reach of the river is

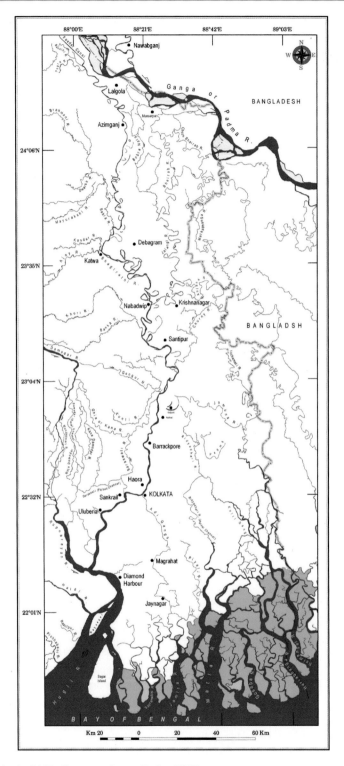

Fig. 6.1 Map of Bhagirathi-Hugli system. *Source* Rudra (2012)

Table 6.1 Freshwater flowing through the Bhagirathi-Hugli River (MCM)

River	Basin area (km²)	Months											
		Jan	Feb	Mar	Apr	May	June	July	Aug	Sep	Oct	Nov	Dec
Bhagirathi (Nabadwip)	22,445	4111	3685	2186	1822	4143	8115	13,074	12,690	10,479	5021	2309	2328
Hugli (Gangasagar)	67,930	4652	4267	2505	2080	5576	11,521	18,528	18,305	14,634	6593	2202	1994

replenished regularly by tidal water. In absence of real time data due to restriction imposed by the Government of India, the mean monthly fresh water flow this river at Nabadwip and at Gangasagar (i.e. outfall) is estimated taking into account rainfall, evapotranspiration, infiltration and storage in the basin (Table 6.1). Notably, tidal flow lager in volume dominates over the fluvial regime in the lower reach of the river.

6.1 Antiquity of the Bhagirathi

The Ganga earlier discharged its water into the Bay of Bengal through many distributaries. It was described as *shatamukhi* (hundred mouths) in the mediaeval Bengali literature. Many scholars opined that the Bhagirathi-Hugli River constituted the oldest outlet of the Ganga water. The decay of Bhagirathi and the eastward flow of the Ganga water is a recent phenomenon (Basu and Chakraborty 1972). The scholars are of the opinion that the diversion of flow took place between twelfth and sixteenth century AD (Majumdar 1942; Hirst 1916). It is further believed that many distributaries like Bhairab, Jalangi, Mathabhanga–Churni, Ichhamati, Gorai-Madhumati were left moribund since the Ganga water escaped through the Padma. This idea was first mooted by Thomas Oldham in 1870. He noted that *'the main water of the Ganges has gradually tracked from the west towards the east, until of late years the larger body of the waters of the Ganges have united with those of the Brahmaputra and have together proceeded to the sea as the Meghna'*.

This opinion was subsequently appropriated by many writers (Bagchi 1944).

Three types of evidences were put forward in favour of the antiquity of the Bhagirathi-Hugli River:

- The great Indian epics of *Ramayana* and *Mahabharata* and classical text like *matsya* and *vayu purana* described the Bhagirathi as the main flow of the Ganga (Ray 1979). But the descriptions in these classical texts are often poetically exaggerated and can hardly be accepted as historical evidences. The human civilization developed earlier in the western side of the delta compared to its eastern counterpart. This can be attributed to the frequent references of the Bhagirathi in the ancient and mediaeval literature.

- The archaeology of the port of Tamralipta which flourished during the millennium of 300 BC to 700 AD has also been referred to as the evidence of antiquity of the Bhagirathi (Rudra 1990). Travellers like Megasthenes (300 BC), Ptolemy (150 AD), Fa-Hien (300 AD), Yuan Chaun (639 AD) described the Bhagirathi and Tamralipta which had trade relation with south-east Asia and the Mediterranean countries (Rudra 1981). However, these descriptions are hardly older than 2300–1300 BP and should be treated as very recent in the geological calendar. The development of a port depends on the economy of the hinterland and might not be always related to the hydrological status of the river.

- The longer southward growth of the delta in the western front is a striking geomorphic phenomenon. Some of the scholars believe that the substantial water and sediment load had been flowing into the sea through the

Hugli estuary during past geological periods. The Meghna estuary, in spite of receiving much greater sediment loads from the Padma and Jamuna, has fallen far behind in the delta building operation from its western counterpart (Majumder 1942).

The Brahmaputra had been flowing through Sylhet basin of Bangladesh till 1830, and much of the sediment carried by the river was trapped in that basin (Coleman 1969). This is one major reason for disproportionate growth of the delta. The coastal tract in West Bengal has retrograded appreciably during the last two hundred years. The possible reasons are absorption of sediment in the reservoirs, wetlands, submarine canyons and erosive wave attack (Bandyopadhyay and Bandyopadhyay 1996). The shape of the delta can further be related to its geology. The Eocene Hinge Zone which extends from south-west to north-east along the subsurface divides the Bengal basin into the stable shelf zone of the north-west and subsiding littoral tract of the south-east. The basement, which receives the sediment load, is tilted eastwards. The depth of sediment strata is much greater along the south eastern part of the delta. Morgan and McIntire (1959) opined that the central part of Bengal basin subsided under the huge load of overlying sediment and a fault might have developed along the course of the Jamuna or Brahmaputra in Bangladesh. This geological characteristic of the delta can conveniently be related to its disproportionate growth, which might not be related to the antiquity of Bhagirathi. The issue of disproportionate growth of delta is also discussed in Chap. 2.

The disproportionate growth of the delta was interpreted largely based on the emotions rather than any rational appreciation, and an old idea has been carried forward by many scholars without proper geomorphological investigation. Fergusson (1863), Oldham (1870) and Reaks (1919) failed to distinguish between the historical and geological timescale. Subsequently, Mukherjee (1938), Majumdar (1942) and Bagchi (1944) were influenced by their predecessors. Only Willcocks (1930) and Chowdhury (1964) expressed different views. The former identified all rivers of central Bengal including the Bhagirathi as irrigation canals excavated by old Hindu rulers. Chowdhury conducted a model study in the laboratory of Cambridge University and concluded that the Padma has been the main outlet of the Ganga.

The cardinal factor controlling the changing course of the Ganga–Brahmaputra delta seems to be the neo-tectonism, especially the Holocene eastward tilt of the basement layer under the weight of the overlying sediment. The eastward migration of the Teesta in 1787 and westward avulsion of the Jamuna in 1830 have been attributed to the subsidence along the line connecting Jalpaiguri and Barisal (Hirst 1916). Morgan and McIntire (1959) identified the subsiding trough along the course of Jamuna in Bangladesh. The Barind and Madhupur tracts were treated as compensatory upheavals.

Since the underlying basement of the GBM delta tilts eastwards, the Ganga water had to comply with the same and flow along Padma channel. It seems to be an absurd proposition that the Ganga water denied this structural control and flowed along the Bhagirathi in the past. The Bhagirathi is the only spill channel of the Ganga which receives tributaries like Pagla–Bansloi, Mayurakshi, Ajay, Damodar, Rupnarayan and Haldi. In compliance with the subsidence of central Bengal, the Padma has deeply incised its valley and consequently its distributaries remain disconnected at their off-takes during lean seasons. The diminishing upstream flow had been the most important cause of the rapid sedimentation in the Bhagirathi-Hugli River affecting the navigation in and out of Kolkata port.

6.2 The Bhairab–Jalangi and the Mathabhanga–Churni

Many distributaries of the Ganga originating from its right bank have been flowing south or south-east barring the Jalangi and the Mathabhanga–Churni which flow south–west and connect the Ganga with the Bhagirathi (Fig. 4.2). Hirst (1916) ascribed this exception to local subsidence. But that idea was not confirmed by any geological investigation, and no change in the tilt of the basin has not been revealed till date. The Jalangi and Churni join the Bhagirathi at Mayapur and Shibpur Ghat (near Payradanga), respectively. The Jalangi River was earlier connected with the Ganga at place also called Jalangi and had been main navigational route to Dhaka (Rennell 1780). But its connection with the Ganga at Jalangi no longer exists, and the Bhairab taking off from the Ganga at Akheriganj maintains the connectivity only during the monsoon months. Both the Jalangi and Mathabhanga–Churni have formed intricate meandering courses. Both the rivers do not receive upstream flow from the Ganga except in the months of August and September when the Ganga becomes bank-full.

The Jalangi and the Mathabhanga–Churni remained delinked from the Ganga since the early twentieth century and now replenished by the monsoon rain and effluent seepage from the groundwater pool (Table 6.2). The presence of many oxbow lakes on the floodplain helps to appreciate how the rivers have changed courses in recent past. The seasonally variable discharge and unconsolidated quaternary sediment layers composing the bank governed formation of intricate meandering. The Gobra Nala which earlier connected the Bhagirathi and the Jalangi, is now almost dry. The total length of the Bhairab–Jalangi from its off-take at Akheriganj to outfall into the Bhagirathi at Nabadwip is 255 km.

The Mathabhanga takes off from the Ganga, at Munshiganj in Kushtia District of Bangladesh. It crosses Indo-Bangladesh border near Gobindapur (Nadia district) and flows about six km to reach Majhdia where it is bifurcated into two branches—the Ichhamati and the Churni. The Ichhamati flows south though Bongaon, Basirhat and Sundarban, and the Churni flows westwards for about 45 km and ultimately joins the River Bhagirathi at Shibpur Ghat (Payradanga). The estimated monthly flow in both Jalangi and Churni at their respective outfall is described in Table 6.2.

6.3 The Changing Off-take of Bhagirathi

The eastward tilt of the delta during the Holocene causing flight of the Ganga towards Bangladesh governed the diminution of flow in the Bhagirathi-Hugli River from its feeder (Morgan and McIntire 1959; Majumder 1942). This deepened the anxiety firstly of the colonial rulers and subsequently that of the Government of India. The difficulty of navigation from the Ganga to the Bhagirathi was experienced by a French traveller named Tavernier, who navigated down the Ganga with M. Bernier on 6 January 1666 and noted '*I left M. Bernier who went to Kasimbazar, and from thence to Hugli by land, because when the river is low one is unable to pass on account of a great bank of sand which is before the town called Soutque*' (Ball 1889). The place spelt as 'Soutque' is now known as Suti and located in Murshidabad district.

Suti was strategically an important location during the mid eighteenth century when the British force was planning to occupy Bengal. Stewart (1813) wrote: '*Seraje ad Dowlah received intelligence, from his spies, that the English were making military preparations; and suspecting that they intended to march towards Moorshudabad, he ordered the division of his army which was encamped near Plassey to be*

Table 6.2 Estimated flow in the Jalangi and the Churni

Basin	Basin area (in km^2)	Estimated flow (MCM)										
		Jan	Feb	Mar	Apr	May	June	July	Aug	Sep	Oct	Nov	Dec
Jalangi	3373	246.	232	190	194	336	69	199	336	488	125	−39	−46
Churni	2598	207.	198	170	172	261	68	63	233	363	75	−30	−34

reinforced; and believing that the English ships of war might proceed up the eastern branch of the Ganga, to the northern point of the Cossimbazar island, and come down the Bhagrutty to Moorshudabad, he commanded immense piles to be driven in the river at Sooty, by which the passage of that river has been rendered merely navigable by boats, and that only during half the year' (Stewart 1813: edition 1903: 593).

The navigation in the Bhagirathi continued to be difficult in the late eighteenth century. James Rennell (1793) in his 'Memoir of a map of Hindoostan' noted *'The two westernmost branches named the Cossimbazar and the Jallinghy river, unite and form what is afterwards named the Hugli river, the only branch of Ganges i.e. commonly navigated by ships. The Cossimbazar river is almost dry from October to May and Jallinghy is in some years unnavigable during two or three driest months'*. The situation remained unchanged compelling India to build the Farakka barrage and excavate a canal joining the Ganga with the Bhagirathi.

Over the preceding centuries, the Ganga–Padma has incised its valley deeply resulting beheading of the Bhagirathi, the Bhairab–Jalangi and the Mathabhanga–Churni and Garai at their respective off-take points. The Ganga flows far below the off-take points in non-monsoon months. With the gradual southward encroachment of the Ganga, the source (or the off-take) of the Bhagirathi has migrated towards the south-east. Notably, acute angular disposition allows easy inflow of water into the distributary but obtuse angle is an unfavourable alignment allowing either no or minimum inflow (Fig. 6.3). In 1796–1797, the Bhagirathi had twin off-take from the Ganga at Dhulian and Suti, respectively (Colebrooke 1801). Between 1824 and 1852, many connecting points between the Ganga and the Bhagirathi were opened and subsequently closed. One of those was located close to Farakka in 1824 (see Fig. 6.2). It shifted to a site 11 km south-east in 1825 further downstream during 1825–1830. In 1847, an inlet was excavated to divert water from the Ganga to the

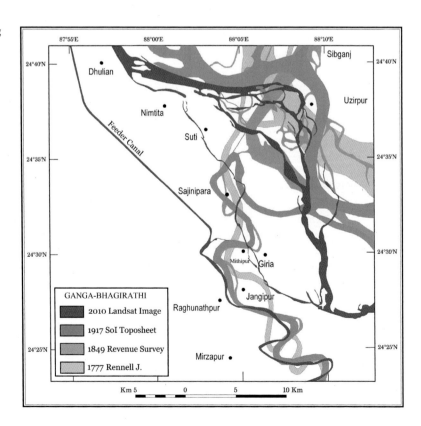

Fig. 6.2 Map of changing off-take of the Bhagirathi. *Source* Rudra (2012)

Bhagirathi but failed. The flood of 1871 opened a new channel to the south-east of the former but it was defunct within a decade. Another attempt to artificially resuscitate the off-take in 1882 was again frustrated (Mitra 1953). The diminishing headwater supply into the Bhagirathi continued, and navigation became an impossible task. In 1917, the Bhagirathi had multiple links with the Ganga at Dhulian, Suti and Giria, respectively. An isthmus covering an area of about 77 km² separated the two rivers between Dhulian and Giria. But the Ganga subsequently migrated westwards, and the two off-takes located at Dhulian and Suti were engulfed. The Bhagirathi continued to remain connected from the Ganga at Mithipur where a defunct obtuse angular off-take still exists (Fig. 6.3).

The Farakka project has connected the Ganga with the Bhagirathi through a feeder canal. The multi-purpose barrage on the Ganga at Farakka, having facility of both road and rail traffic, was designed to divert water (40,000 cusec or 1133 m³ s⁻¹) from the barrage pond into the Bhagirathi. The barrage site was chosen at a nodal point where the bank was found resistant to erosion. A 38-km-long Ganga–Bhagirathi canal was excavated to induce 1133 m³ s⁻¹, water into the latter (Fig. 6.3). A smaller barrage was constructed at Jangipur to stop the escape of water back to the Ganga and to ensure flow of water towards the port of Kolkata (Basu 1982).

An afflux bund between Farakka and Nimsarai was built across moribund distributaries, the Pagla and the *Chhoto* Bhagirathi, so that the water from the Farakka pond does not get the opportunity of outflanking the structure across the Ganga (Fig. 4.7). The engineers apprehended that the Ganga might open a wider route through the moribund channels and bypass the Farakka barrage during a devastating flood. The intake point of the Pagla was closed by earthen embankment, and the same of the *Chhoto* Bhagirathi was locked with a head regulator. The Bhagirathi between Jangipur and Nabadwip which was fordable before 1133 m³ s⁻¹ of water was induced in 1975. The water level of the Bhagirathi was raised due to induced water, and four tributaries, namely Bansloi, Pagla, Mayurakshi, Ajay in the west bank and the Jalangi and Churni in the east bank, faced drainage congestion at their respective outfalls. Since the tributaries find it difficult to discharge water into the Bhagirathi under altered hydrological condition, the recurrent floods have been common place.

6.4 Meandering Channels

The rivers, especially the distributaries, in the deltaic tract meander and change their courses in such a fashion that it may appear as abnormal (Hirst 1916). The geometry of meander changes during rainy season when the historically highest discharge flows through the river. The bend of a river gradually takes an extreme shape, and the flowing water finally cut through the neck forming abandoned meander loop. The annual peak discharge and stratigraphy of bank are two important factors governing bank erosion and meander migration. Notably, each river has a meander belt where it moves laterally (Fig. 6.4).

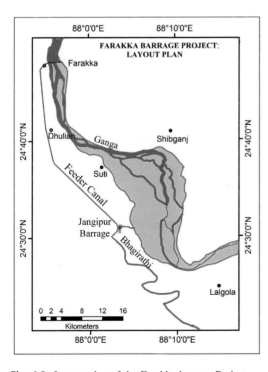

Fig. 6.3 Layout plan of the Farakka barrage Project

The Bhagirathi is very active between Jangipur and Nabadwip. In southern tidal regime, the tendency of the river to erode its bank is much less as finer clay–silt layer forming bank resist erosion.

The floodplain of the Bhagirathi is incidentally bordered by railway and highway on both banks. The bends formed by the Bhagirathi are variable in magnitude. The largest bend is formed near Katwa at the outfall of the Ajay River and remained unchanged for many years (Fig. 6.5). The distance from the crest of one bend to other may vary between 2.3 and 9.5 km.

Since 1133 $m^3\ s^{-1}$ of water was diverted from the Ganga and was induced into the Bhagirathi, it started to adjust itself with altered hydraulic condition. The Bhagirathi cut the necks of meanders and adopted some short-cut route leaving behind oxbow lakes on the floodplain. In 1984, the Bhagirathi opened new route and its course near Dadpur was reduced by two kilometres. Its old course still exists on eastern bank. The river has formed intricate meandering down the outfalls of the Mayurakshi and the Ajay. The discharge and sediment load contributed by the two tributaries compel have had appreciable impact on the meander geometry. The comparison of multi-dated maps helps to visualize how the Bhagirathi migrated laterally since 1917–2010 (Fig. 6.5).

The river alters its course frequently in some vulnerable stretches. In 1989, the Bhagirathi opened a new route just upstream of Nabadwip and created an oxbow lake on the western bank (Fig. 6.6). It attracts many migratory birds during the winter and has developed as a tourist spot. A similar event happened in 1994 near Santipur where the river left its former meander loop on the east bank. The west bank of the Bhagirathi has an eastward slope and relatively higher than the eastern floodplain. The tributaries draining western plateau ultimately drain to the Bhagirathi. The water level exceeding the threshold limit often submerges the adjoining floodplain and breaches the embankment. Even the National Highway 34 and the railway were breached many times. In September 2000, the Bhagirathi flowed over the eastern bank and virtually avulsed

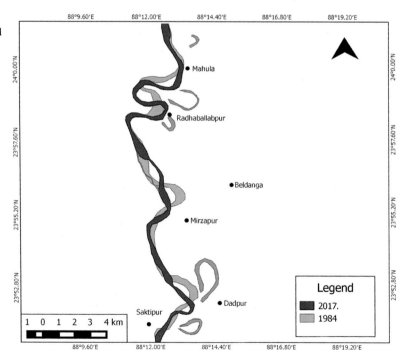

Fig. 6.4 Meander cut-off between Radhaballabpur and Dadpur

Fig. 6.5 Great bend and intricate meandering near Katwa. *Source* Rudra (2012)

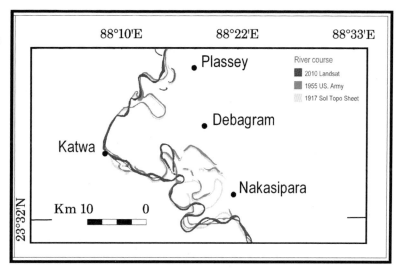

Fig. 6.6 Map showing lateral oscillation between Nabadwip and Santipur. *Source* Rudra (2012)

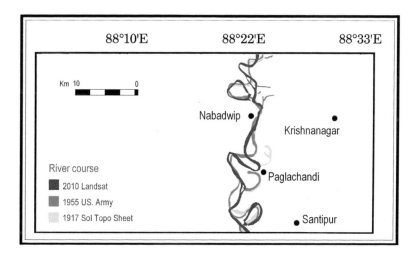

through the Gobra River causing loss of crops and death of many (Chapman and Rudra 2016).

The flood and river bank erosion are two inherent problems of Bengal. The people of rural Bengal always prefer to live close to rivers. The density of population along the bank has increased many times in recent years due to high birth rate and large-scale population migration. So displacement of large number of people due to flood and erosion is common in Bengal.

6.5 Dying Rivers

The Bhagirathi–Hugli River flows southwards for about 300 km between Jangipur and Tribeni and released three channels reaching at Tribeni during the mediaeval period (Fig. 6.7). The place names in Bengal are often linked with its rivers as Tribeni means three braids of hair which resemble three channels branching off from a common

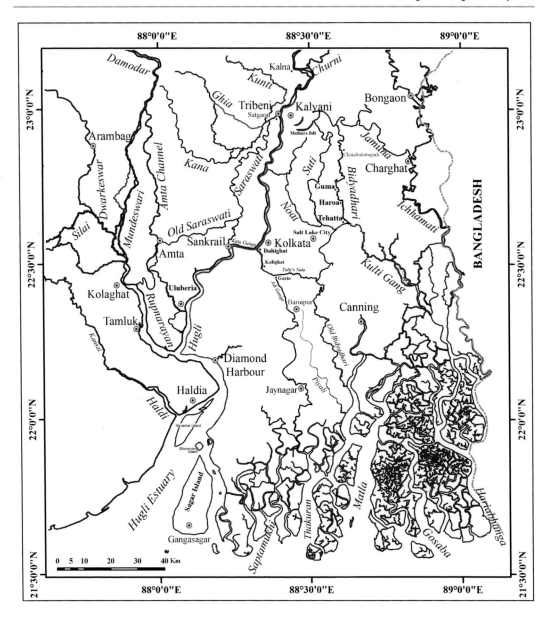

Fig. 6.7 Map showing lower Bhagirathi distributaries. *Source* Rudra (2014)

point. The distributaries which flowed towards three different directions are the Saraswati, the Hugli and the Jamuna. The Bidyadhari earlier took off from the Hugli River near Kalyani but was beheaded due to shifting of the latter. The oxbow lake which encircles Kalyani town was part the Hugli River. The Adi Ganga is treated as a sacred branch of the Ganga because the channel originates at Dhahighat and flows along the famous Hindu temple of Kalighat. Unfortunately, 57 drains discharge wastewater of the southern part of city into this channel and pose serious threat to public health. A heritage river has been gradually converted into a wastewater outlet. It is a classic example to show how unplanned urbanization may lead to pollution of a river.

6.6 The Saraswati River

The Saraswati is a moribund distributary of the Hugli River having its source at Tribeni and outfall into the same river at Sankrail. The river travels 77 km and remains almost dry during lean months. The Satgaon, which developed on the left bank of this river, was an important port of Bengal during mediaeval period.

The Saraswati River and port of Satgaon were intertwined in many historical essays. Satgaon was described as *porto pequeno* meaning the 'little port' by the sailors who came from Portugal in the sixteenth century (Sarkar 1973). The seat of administration of the southern Bengal was taken from Tribeni to Satgaon in 1329 AD, and the latter gradually emerged as a port town (Rudra 1990). The flow of water through the Saraswati diminished during the second half of the sixteenth century, and the sailors found it difficult to approach Satgaon and the port gradually declined. It is important to note that the Saraswati in its lower reach flowed through a different course till the seventh century AD and discharged through the Rupnarayan estuary. Tamralipta the oldest port of Bengal was located on the right bank of the Rupnarayan River. The link between the Rupnarayan and Saraswati was subsequently disconnected but the estuary was called 'old Ganges' by the local people even after couple of centuries. Rennell (1793) in his *Memoir of a Map of Hindoostan or the Mughal Empire* noted '*Satgong or Satagong, now an inconsiderable village on small creek of the Hugli river, about four miles to the north–west of Hugli was in 1566 and probably later a large trading city, in which European traders had their factories in Bengal. At that time Satgong was capable of bearing small vessels, and I suspect that its then course passing Satgong was by way of Adampour (West Chanditala), Omptah (Amta) and Tamlook (Tamluk), and that the river called old Ganges was part of its course, and received that name when the circumstances of the change was fresh in the memory of people. The appearance of the country between Satgong and Tamlook countenances such opinion*'.

The Tamralipta (now known as Tamluk) was an important port in the Indo-Roman coastal trade during 300 BC–700 AD. It was also connected with the south-east Asian countries. The relics of the Tamralipta were found buried under sedimentary layers of the right bank of the Rupnarayan river which is a tributary to the Hugli River (Fig. 6.7). In many travelogues and classical texts, Tamralipta was described as an important trading centres located on the bank of the Ganga not on the Rupnarayan (Rudra 1981). This description seems to be wrong unless one appreciates how the water of the old course of the Saraswati served the port of Tamralipta. Notably, a rainfed tributary like Rupnarayan cannot facilitate navigation in and out of an international port. The Rupnarayan carried less volume of water in the past, as the Damodar which now discharges a bulk of water through the Mundeswari into the Rupnarayan flowed through two other branches joining the Bhagirathi-Hugli one eastwards near Kalna and other near Uluberia (Sen 1968). The Rupnarayan estuary which had been one of the outlets of the Bhagirathi-Hugli River now acts as a tributary (Rudra 1990).

The branch of Saraswati connecting it with the Rupnarayan has been identified on recent satellite image, and a map is prepared (Fig. 6.7). This channel was also described by Fergusson (1863). The decay of channel was caused by eastward flight of the flow from its off-take at Tribeni and rapid sedimentation. The Saraswati withdrew its allegiance to the Rupnarayan since 700 AD and migrated eastwards opening a new outlet along Sankrail leading to decline of the port of Tamralipta (Rudra 1990). The present Saraswati is a linear pool of stagnant swamp having feeble flow at both ends.

6.7 The Bidyadhari–Sunti–Noai System

The Bidyadhari has its source near the Mathura *Bil* which appears as an abandoned meander cut-off of the Bhagirathi located close to Kalyani.

It is also known as Nonagang between Kalyani and Guma and thereafter as the Bidyadhari. The local people often call it as Sealdahgang and the Haruagang (Mitra 1995). Earlier the river had two branches being bifurcated at Tehatta. One branch flowed through present Salt Lake City and discharged into the Bay of Bengal through the Matla estuary till the 1950s. The second branch, locally known as Kulti Gang, opened its outlet through the Haribhanga estuary. In the decade of 1770, Colonel Tolly excavated a canal from Garia to Samukpota which connected the Adi Ganga with the western branch of the Bidyadhari (Fig. 6.7). That canal gradually turned into a sewage outlet and decayed. Thus, western branch of the Bidyadhari was dissociated from the Matla (Maitra 1969). Claudius Ptolemy (150 AD), the famous Greek cartographer, described five estuaries of the Ganga, namely *Kambyson*, *Mega*, *Kamberikhon*, *Pseudostomon* and *Antibole*. The Bidyadhari discharged into the sea through the 'Mega' which is now named as the Haribhanga estuary (Rudra 1981). This was an important navigational route for Indo-Roman coastal trade during 300 BC–500 AD, and the business was conducted from the port of Chandraketugarh which was contemporaneous to port of Tamralipta and stood on bank of Bidyadhari (Ray 1979). The flow of water in the Bidyadhari reduced after 500 AD when the Bhagirathi shifted westward near Kalyani leaving behind Mathura *Bil* as an oxbow lake. The catchment of Bidyadhari along with the Noai and the Sunti, two other moribund channels, is most densely populated. The fluvial regime has changed, and flow is intercepted by human intervention at many places to facilitate irrigation. Even the bed of the rivers is converted into agricultural land at places. The urban centres developed along the bank use the channels for discharging wastewater and converted them into swamps at places.

6.8 The Jamuna and the Ichhamati

The Jamuna (common name with river in Bangladesh) is a moribund channel which connected the Hugli River and Ichhamati and was presumably opened since the desiccation of the Bidyadhari. The tides advancing northwards through formerly active Saraswati and the Bidyadhari channels would meet at Tribeni and accelerated process of sedimentation leading to formation of a mid-channel bar. The southward flow was obstructed compelling the opening of a new channel towards east (Rudra 1990). The Jamuna presently having no link with the Hugli River but it is extremely sinuous course is traceable for about 66 km down to Charghat where it joins the Ichhamati. Both the Jamuna and the Ichhamati were navigable for large naval ship in the early seventeenth century, and a local king fought a heroic naval war against the external force in 1612 AD at confluence of two rivers (Sarkar 1973). Both the channels have gone dry during subsequent centuries, and it is often difficult for a country boat to sail.

The Jamuna was first described as an active channel in a literary work called *Pavandutam* written by Dhoyee in 1175 AD (Chakraborty 1924). Bipradas Piplai (1495 AD; quoted in Ray 1979) in his treatise *Manasamangla* described it as a *bishál* or a mighty river. This description seems to a poetic exaggeration (Ray 1979) but Hunter (1875) described the Jamuna as a navigable river which could facilitate ply of large trading boats. However, Rennell (1780), while surveying Bengal in the second half of the eighteenth century, found the Jamuna as an insignificant channel and drew it accordingly in his 'A Bengal Atlas'.

The Ichhamati is another beheaded distributary. The Mathabhanga is bifurcated at Majhdia, in Nadia district, and releases the Churni to the west and Ichhamati to the south. The latter flows about 240 km touching several towns along the Indo-Bangladesh border and finally discharges into the sea through the Haribhanga estuary in Sundarban. The Ichhamati does not receive upstream flow from the Mathabhanga except during the flood but the latter was active in 1795 AD and maintained its status even subsequently (Garrett 1910). The Mathabhanga was reportedly more active than the Bhagirathi in the early nineteenth century and was an important navigational route between Kolkata and Dhaka. The upper Ichhamati is now virtually a stagnant pool of water and full of water hyacinth, but its lower tidal regime looks better due to regular swelling of water.

6.9 The Adi Ganga

The Adi Ganga is moribund distributary of the Hugli River and originates at Dhahighat where the former takes sharp bend (Fig. 6.7). The course of river was aligned through Khiddirpur Alipur Kalighat, Tollygunge, Garia, Baruipur and Jaynagar and finally reached the Bay of Bengal through the Saptamukhi estuary. There is a popular misconception that the Adi Ganga once cut across Sagar Island to reach the Bay of Bengal. But fact is the tidal channel which bifurcated the island earlier and subsequently went dry had never been connected with the Adi Ganga. Some mediaeval Bengali texts like the *Mangalkabya* and the *Chaitanyabhagabat* referred the Adi Ganga as a popular trade route (Rudra 1990). This channel is treated as very sacred by the Hindus because of the location of famous Kalighat temple along its bank. Since an elevated corridor was built over the river to extend metro railway from Tollygunge to Garia, flow was hindered. Further south, rapid urbanization has encroached upon the river at Garia, Sonarpour and Baruipur. The Piyali connected the Adi Ganga and the Bidyadhari till the late nineteenth century but now lies in moribund condition (Hunter 1875). The old course of the Adi Ganga between Jaynagar to the estuary is difficult to identify in both maps and ground. The maps are prepared by Rennell (1780), Tassin (1841), and satellite images help to identify the northern part of the Adi Ganga beyond Jaynagar but its course further south is not traceable. Bandyopadhyay (1996) explored the lost course of Adi Ganga from literary and cartographic evidences and opined that the channel might have flowed to the Bay of Bengal through the Saptamukhi or the Thakuran estuary. Rudra (1986) identified the former as the outlet of the Adi Ganga.

The decay of the Adi Ganga was aggravated firstly due to artificial diversion of the water from Dhahighat to Sankrail through an excavated canal called *Kata* Ganga in 1739–56. It was reportedly done by Alibordi Khan, the then Nawab of Bengal; and secondly, the situation deteriorated further when the Tolly's *Nala* connecting the Adi Ganga with the old Bidyadhari was excavated during 1775–77 (Rudra 2008). The Adi Ganga virtually died in first decade of the twenty-first century when the metro railway between Tollygunge and Garia was extended putting more than 300 piers on the river bed. The river in this stretch now carries some polluted water discharged through 57 outlets. In spite of tidal water regularly reaching up to Kalighat temple and diluting pollution load, the heritage river looks like drain carrying urban wastewater.

6.10 The Hugli Estuary

The Hugli estuary is one of the major outlets of the Ganga basin. It is a funnel-shaped mouth which allows tidal water to invade the river. The saline water entering from the sea is mixed with freshwater drained from the catchment in this estuary (Pethick 2000). The port of Kolkata is located about 150 km inland from the sea. The port of Kolkata was built in the late eighteenth century under colonial rule but the navigation in this river was hampered due to sedimentation and shallow depth. The merchant ship could ply in the river only during high tide because fourteen bars lying below the low-tide level impeded navigation in the channel leading to Kolkata port. The tide and freshwater dynamics in the Hugli estuary is extremely complex, and tidal range varies widely between sea front (Sagar) to Kolkata (Garden Reach). The tidal swelling may be more than five metres at Sagar and six metres at Kolkata when the river achieves highest water level (Table 6.3). The Hugli estuary is unique compared to twelve other estuaries of the GBM delta. The Bhagirathi-Hugli River flows from the north to the south on its way to the sea; on the contrary, all other distributaries flow south-easterly. The funnel-shaped Hugli estuary allows huge sediment-laden water to penetrate and choke the channel (Fig 6.8).

Table 6.3 Tidal swelling in the Hugli estuary (in metre)

Locations/level	Sagar	Ganga	Haldia	Diamond Harbour	Mayapur	Garden Reach
Highest high water	6.66	7.25	7.26	7.35	7.10	7.70
Mean high water	4.64	4.93	5.01	5.24	4.83	4.88
Mean low water	1.51	1.39	1.34	1.47	1.28	1.68
Chart datum	0.46 m below Khidirpur old dock site					
Lowest low water	−0.21	+0.03	−0.07	+0.08	+0.03	+0.14

Source Kolkata Port Trust (2016)

Fig. 6.8 Changing Hugli estuary

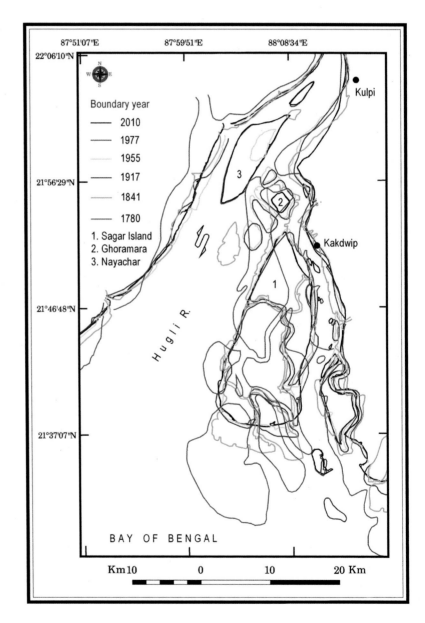

The Farakka barrage was commissioned in 1975, and it was expected that the induced water from barrage would help to improve navigation facility. But additional flow diverted from the Ganga was subsequently found incapable to stop sedimentation in the estuary. The tidal bores invade the estuary frequently. The tide-velocity asymmetry largely accelerates the process sedimentation. The penetrating tide has average velocity of 2–3 m/s^{-1} and tends to deflect towards east or left bank of the river governed by the Coriolis force. The water level swells from 1.51 to 6.66 m at Gangasagar and from 1.47 to 7.35 m at Diamond Harbour within 4 h (Table 6.3). The inflowing tidal water goes back to the sea during ebb tide at a slower velocity (<1 m/s^{-1}) in succeeding 8 h. A substantial portion of the sediment load carried by the high tide is deposited on the river bed. Hirst (1916) noted that the *'forces controlling it are so powerful that any artificial interference would be futile'*. Since mangroves were cleared to facilitate human settlement and agriculture during the colonial rule, the coast was exposed to wave attack (WWF India 2010). The sea has advanced at least 12 km since 1780 and continues to encroach further inland (Fig. 6.8). The famous temple located at southern face of Sagar Island was shifted many times under threat of coastal erosion.

6.11 Suspended Load

The Hugli estuary has been in alarming state of decay making the navigation of seagoing vessels difficult since the late eighteenth century when Kolkata emerged as a port town. While suspended load in the river increased with expansion of agriculture in the catchment, its ability to carry the same down the estuary has been reduced due to diminishing freshwater flow (Rudra 2014). When five reservoirs were built across the Damodar and its tributaries and the Mayurakshi and the Kansai were tamed with the objective of impounding monsoon flow to facilitate irrigation, the highest flow of the Bhagirathi-Hugli was diminished. Consequently, the capacity of the river to flush the sediment load into the deeper estuary was reduced. The dredging operation of navigation channel started since 1820 and continues even after commissioning of the Farakka barrage Project (Ghosh 1972). But continuous dredging failed to ensure the draft required for movement of the large ship (Sanyal and Chakraborty 1995). The unpublished reports of Kolkata Port Trust revealed that 2.96×10^6 tons of sediment from the channel and 23.22×10^6 tons of the bank forming materials were flushed downstream during the period 1976–1991 (KPT 1999–2000). While 4.31×10^6 tons of sediment load drained annually into the Bhagirathi-Hugli from the western undulating terrain, the eastern floodplain contributes 0.18×10^6 tons every year. Notably, the reservoirs built across the western tributaries to the Bhagirathi-Hugli River trapped and hindered downstream movement of sediment load.

The Ganga–Bhagirathi feeder canal diverts transfers sediment load weighing 0.14×10^6 tons/year. This relatively silt-free water quickly erodes riverbed and bank and carries sediment load downstream. The suspended load weighing 26.93×10^6 ton moves to and fro in the estuary with tidal water (Rudra 2014). It has been observed by the hydraulic study department of Kolkata Port Trust that 11.02 million m^3 of sediment load is transferred annually from non-tidal regime into the estuary but the volume of the dredged material was in the order of 0.29 million m^3. Thus, 10.73 million m^3 of sediment is either precipitated on the bed or remains in movement with water.

When port of Kolkata became inaccessible for large vessels in the late twentieth century, Haldia (Fig. 6.7) was built as a substitute port. But the problem of inadequate draught for the movement large ship continues. The plan to build a port further south or a deep-sea anchorage is now under consideration. The unabated sedimentation leading to formation many underwater shoals, sinuous navigation channel and feeble freshwater flow are the constraints for navigation in this river. Notably, the port of Kolkata was developed to serve colonial interests when light wooden ship had been plying and the depth had never been adequate for large vessels which dominate international trade today. The uninterrupted sedimentation had been an age-old problem, and consequently the location of the port shifted southwards from Kolkata to Haldia.

The fluvio-marine processes operating in the estuary are complex. The freshwater supply down the Bhagirathi-Hugli River has diminished appreciably with expansion of irrigation in the catchment of the river. The water diverted through the Ganga–Bhagirathi feeder canal has rejuvenated the non-tidal regime and reduced salinity of flowing water. But the target of the Farakka barrage Project was partially achieved. In fact, the freshwater flow was inadequate to thwart huge the volume of tidal flow from the Bay of Bengal and struggle for safe navigation continues.

References

Bagchi K (1944) The Ganges Delta. University of Calcutta, Kolkata, pp 50–70
Ball V (1889) Travels in India by Jean Baptista Tavernier, Macmillan and Company, London 1:125–126
Bandyopadhyay S (1996) Location of the Adi Ganga palaeochannel, South 24 Parganas, West Bengal: a review. Geogr Rev India 58:93–109
Bandyopadhyay S, Bandyopadhyay MK (1996) Retrogradation of the western Ganga-Brahmaputra delta, India and Bangladesh, possible reasons. Nat Geogr 31: 105–128
Basu SK (1982) A geotechnical assessment of the Farakka Barrage Project, Murshidabad and Maldah Districts, West Bengal. Bull Geol Surv India 47:2–3
Basu SR, Chakraborty S (1972) Some considerations over the decay of the Bhagirathi drainage system. In: Bagchi K (ed) The Bhagirathi-Hooghly Basin, proceedings of interdisciplinary symposium. University of Calcutta (Kolkata), pp 57–77
Chakraborty C (ed) (1924) Pavandutam of Dhoyee (in Sanskrit), Sahitya Parishad book no. 12. Kolkata, pp 1–60
Chapman GP, Rudra K (2016) Water as Foe, Water as Friend. J South Asian Develop 2(1):19–49
Chowdhury MI (1964) On the gradual shifting of the Ganges from west to east in the delta building operation. In: Proceedings of the symposium on scientific problems of the humid tropical zone deltas and their implications, Dacca, UNESCO, 35–40
Colebrooke RH (1801) On the courses of the Ganges through Bengal. Asiatic Res Asiatic Soc Kolkata 7:1–31
Coleman JM (1969) Brahmaputra: channel processes and sedimentation. Sed Geol 3:1–131

Fergusson J (1863) Recent changes in the delta of the Ganges. Q J Geol Soc (of London) 19:321–354
Garrett JHE (1910) Bengal District Gazetteer, Nadia. Reprinted by West Bengal District Gazetteer, Government of West Benal in 2001. Kolkata, pp 5–24
Ghosh HP (1972) History, development and problems of dredging in the Hooghly River. In: Bagchi K (ed) The Bhagirathi-Hooghly Basin, proceedings of interdisciplinary symposium, University of Calcutta (Kolkata), pp 174–189
Hirst FC (1916) Report on the Nadia Rivers. Reprinted by Gazetteer Department, Government of West Bengal, Kolkata, pp 3–61
Hunter WW (1875) A statistical account of Bengal. Reprinted by Gazetteer Department, Government of West Bengal, Kolkata, pp 9–13
Kolkata Port Trust (2000) Unpublished Annual Report of Hydraulic Study Department, Kolkata
Kolkata Port Trust (2016) Tide tables for Hugli River. Kolkata
Majumder SC (1942) Rivers of Bengal Delta. Reprinted by Gazetteer Department, Government of West Bengal, Kolkata, pp 7–102
Maitra B (1969) River systems of West Bengal. River Research Institute, Government of West Bengal. Kolkata, pp 1–125
Mitra A (1953) History of the mouth of the Bhagirathi River 1781–1925, Selection of Records of the Government of Bengal relating to Nadia Rivers (from 1848–1926). Reprinted in District Census Handbooks/ Murshidabad 1961, Appendix vi, pp 111–133
Mitra S (1995) Uttar Chabbis Parganar Nadi O Prachin Savyata (in Bengali). Jana Sanhati Kendra, Madhyamgram, 24 Parganas, West Bengal, pp 1–20
Morgan JP, McIntire WG (1959) Quaternary geology of the Bengal basin. East Pak India Bull Geol Soc Am 70:319–342
Mukherjee RK (1938) The Changing Face of Bengal/A Study of Riverine Economy. University of Calcutta
Oldham T (1870) Address of the President. In: Proceedings, Asiatic Society of Bengal, Kolkata, pp 40–52
Pethick John (2000) An introduction to coastal geomorphology. Arnold, London, pp 167–187
Ray NR (1979) Tamralipta and Gange: two port cities of ancient Bengal and connected considerations. Geogr Rev India 41:205–222
Reaks HG (1919) Report on the physical and hydraulic characteristics of the rivers of the delta. Reprinted in Rivers of Bengal, vol. II, West Bengal District Gazetteer (2001)
Rennell J (1780) A Bengal atlas containing maps of the theatre of war and commerce on that side of Hindoostan. London (edited by Kalyan Rudra (2016) and reprinted by Sahitya Samasad, Kolkata
Rennell J (1793) Memoir of a Map of Hindoostan or the Mugul Empire. Reprinted by Editions Indian in 1976, Kolkata, pp 148–150
Rennell J (2016) The journal of James Rennell. Edited by La Touche THD (1910); reprinted in volume II of A Bengal Atlas containing maps of the theatre of war

References

and commerce on that side of Hindoostan. London edited by Kalyan Rudra (2016); published by Sahitya Samasad, Kolkata, pp 13–17

Rudra K (1981) Identification of the Ancient mouths of the Ganga as described by Ptolemy. Geogr Rev India 43:97–104

Rudra K (1986) The history of development of the Bhagirathi-Hugli River and some connected consideration, Unpublished Ph.D. thesis, Department of Geography, University of Calcutta

Rudra K (1990) Tamralipta and its locational problem. In: Ray A, Mukherjee S (eds) Historical archaeology of India. Books and Books, New Delhi, pp 245–254

Rudra K (2008) Banglar Nadikatha (in Bengali). Sahitya Samsad, Kolkata, pp 1–116

Rudra K (2012) Atlas of changing river courses in West Bengal. Sea Explorers' Institute, Kolkata

Rudra K (2014) Changing river courses in the western part of the Ganga-Brahmaputra Delta. Geomorpholgy 227:87–100

Sanyal T, Chakraborty AK (1995) Farakka barrage project: promises and achievements. In: Chakraborty SC (ed) 125 years of Kolkata port trust, commemorative volume. Kolkata Port Trust, Kolkata, pp 55–58

Sarkar JN (1973) The history of Bengal/Muslim period (1200–1757). Academica Asiatica, Patna, pp 318–319

Sen S (1968) Major changes in river courses in recent history. In: Law BC (ed) Mountains and rivers of India. National Committee of Geography, Calcutta, pp 211–220

Stewart C (1813, references above to 1903 reprint) The History of Bengal: from the first Muhammeden Invasion until the Virtual Conquest of that Country by the English A.D. 1757, Calcutta

Tassin JBB (1841) A New Bengal Atlas, Calcutta

Willcocks W (1930) The ancient system of irrigation in Bengal. Calcutta University

WWF India (2010) Sundarban: future imperfect, climate adaptation report. Kolkata, pp 3–28

The Western Tributaries to the Bhagirathi–Hugli River

Abstract

The Bhagirathi–Hugli River is a branch of the Ganga which receives seven tributaries and having a combined catchment area covering 66,000 km². These tributaries are replenished by rainwater during the four monsoon months and remain almost dry during the non-monsoon months. But the lower catchments of these rivers are plains and prone to recurrent floods. Even after embanking and building of reservoirs across many rivers, the floods have not been regulated. The Damodar which was earlier described as 'Sorrow of Bengal' and tamed both by embankment and dam/reservoir continues to imperil its lower catchment by recurrent floods.

The Bhagirathi–Hugli River and the Ganga are artificially linked through a 38 km long feeder canal (Fig. 7.1). The seven major tributaries drain western uplands of the Bhagirathi–Hugli basin covering an area of about 66,000 km². These tributaries are of different length and magnitude and exclusively rainfed. Since rainfall is mostly concentrated in the four monsoon months, the rivers reach the bank-full stage only during August and September and often spill over the floodplain. But these rivers become feeble during lean months. Each of these basins is divided into two topographic units (a) the undulating terrain and (b) lower plain. The lower parts of catchments are flood prone. Two topographic units are differentiated by 18 m contour line (Bagchi and Mukherjee 1979). The larger magnitude of the upper unit makes the lower one flood prone. Broadly speaking, these rivers are east flowing and tend to flow south-east in the lower reach. Debouching on the plains, some larger rivers have changed their courses. Such change is most significant in lower Damodar Valley (Rudra 2008).

7.1 The Bansloi

Taken from the north, Bansloi is the first tributary of the Bhagirathi, which has its source at Ramgarh near Rajmahal Hills and outfall at Jangipur. The river drains an area about 1176 km² and its length is 112 km from the source to the mouth.

The five tributaries to Bansloi are Bhorsha, Karjor, Ikhri, Sukhra and Danro. Since the river is exclusively rainfed, it goes dry during lean season. Table 7.1 describes the mean monthly flow generated in this basin. The discharge that runs off the basin is highest during the July when about 394 MCM water flows. The lowest flow is estimated in the month of December when evapotranspiration exceeds rainfall. Since the induction of 1132 cumec of water into the Bhagirathi from

Fig. 7.1 Map showing drainage congestion in Bansloi and Pagla basins

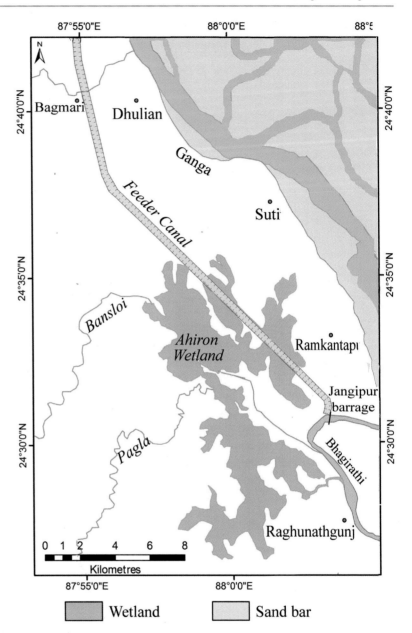

the Farakka barrage, the water level of the Bhagirathi at Jangipur was raised to about 5 m causing severe drainage congestion in lower Bansloi and Pagla basin. An area of 50 km² was perpetually waterlogged which is locally known as *Ahiraner jala*. The National Highway no. 34 and the railway connecting Jangipur with Farakka also act as embankments and aggravate the problem of waterlogging. The subsequent construction of two sluices at the outfalls of both the Bansloi and the Pagla did not produce any effective improvement of drainage congestion.

Table 7.1 Estimated discharges in the Bagmari and the Bansloi Rivers

Basin	Basin area (km^2)	Estimated flow (MCM)											
		Jan	Feb	Mar	Apr	May	June	July	Aug	Sep	Oct	Nov	Dec
Bagmari	671	36	31	19	16	40	29	91	122	141	22	−10	−10
Bansloi	1176	49	36	4	−5	68	223	394	377	374	74	−26	−27

7.2 The Pagla River

The Pagla has a catchment of about 628 km^2 of which 377 km^2 falls in West Bengal and 251 km^2 in Jharkhand. The estimated maximum discharge in this river has been observed in the month of July when 138 MCM of water flows through the river. The river goes dry in the month of December (Table 7.2). This channel originates from Masina in Santhal Parganas of Jharkhand where the river is known as Surjudi. The river takes the name Pagla further downstream where a minor tributary joins it. Suri is the most important tributary which joins Pagla at Kalikapur in Birbhum District. The river has a length of 80 km and its lower 45 km is in Birbhum and Murshidabad Districts of West Bengal (Fig. 7.2).

Prior to the excavation of Farakka–Jangipur Feeder Canal, Bagmari River, which drains a basin area of 671 km^2 used to discharge into the Ganga. About 627 km^2 of Bagmari basin is in Jharkhand while 44 km^2 is in West Bengal. A siphon was constructed under the Farakka Feeder Canal to render the river an outlet towards the Ganga. This was subsequently choked aggravating the problem of drainage congestion. Since this river is also exclusively rainfed, 80% of the annual flow of water passes during the monsoon months. In September 2000, an unprecedented cloudburst generated a huge discharge in the Bagmari catchment and that water escaped towards the Bhagirathi through the feeder canal. The volume of water flowing through the Bhagirathi at Jangipur was more than 3000 million m^3 and that caused breach of the left embankment of the Bhagirathi at Kalukhali and flowed through the moribund channel of Gobra *Nala*; and Murshidabad and Nadia Districts experienced the most disastrous flood ever known to history (Chapman and Rudra 2007).

7.3 The Mayurakshi

The Mayurakshi is the third important western tributary of the Bhagirathi. It originates from Trikut Hills in Deoghar District of Jharkhand at an altitude of 610 m and finally discharges into the Bhagirathi near Kalyanpur. The major tributaries draining the upper Mayurakshi catchment are Dhobbi, Pussaro, Bhamri and Tepra. These tributaries originate from the uplands of Santhal Parganas. The river runs from the west to the south-east through the Santhal Parganas and after combining with the Nunbil and the Siddheshwari at Sadipur, it enters the Birbhum District of West Bengal (GoWB 1989). The two important left bank tributaries are the Dwarka and the Bhrahmani which are combined together at Natunhat located in a depression locally known as Hijal *Bill*. Two other right bank tributaries Kopai and Bakreshwar are combined to form the Kuea River which ultimately joins the Mayurakshi at a place called Sunoti. All tributaries draining into the Hijal *Bill*, finally bifurcated into two outlets,

Table 7.2 Estimated flow in the Pagla River

Basin	Basin area (km^2)	Estimated flow (MCM)											
		Jan	Feb	Mar	Apr	May	June	July	Aug	Sep	Oct	Nov	Dec
Pagla	628	14	11	1	−2	26	82	138	129	111	29	−8	−9

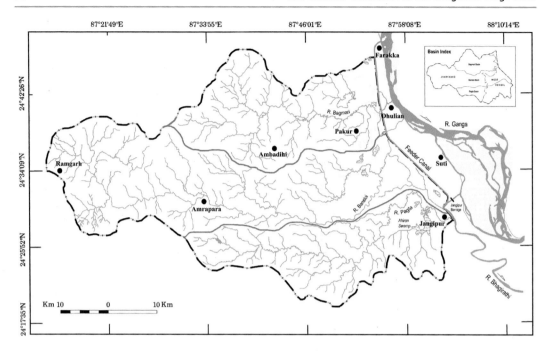

Fig. 7.2 The Bagmari, the Bansloi and the Pagla basins

namely Uttarasan and Babla towards the Bhagirathi River (Fig. 7.3). The Uttarasan often experiences a reversal of flow when the Bhagirathi is in spate during the peak of monsoons. The principal flow of the river passes through Babla which joins the Bhagirathi at Kalyanpur. The total catchment area of Mayurakshi is about 9345 km^2.

The maximum discharge (2031 MCM) flows in the month of July while minimum flow is recorded in December (Table 7.3). The 71% of the annual flow is generated in the four monsoon months. The sediment load carried by the river annually about three million tons. The lower Mayurakshi catchment is an area of acute drainage congestion. The Mayurakshi River was dammed at Massanjor in mid-1950s, and a barrage was also constructed at Tilpara to facilitate irrigation. A detailed account of the Mayurakshi Barrage Project is given in Chap. 9.

The lower course of the Mayurakshi has changed since the late eighteenth century when it used to join the Bhagirathi at place then called Jumjumcoly. The Kopai (described as Cupi by James Rennell (1780) in his Atlas), a right bank tributary of the Mayurakshi, discharged independently into the Bhagirathi at Idilpur. There was a feeble channel connecting these two rivers. It is not known till date that when the Mayurakshi turned southwards and adopted lower Kopai as the route to the Bhagirathi.

7.4 The Ajay

The Ajay River originates from the uplands of Hazaribag and flows about 276 km before it joins Bhagirathi at Katwa. It drains a catchment of 6074 km^2 which is shared by both Jharkhand and West Bengal. About 58% of the basin (3526 km^2) lies in Jharkhand and remaining 42% or 2548 km^2 goes to West Bengal. The Kunur is the most important right bank tributary which originates near Ranigunj and flows through Ausgram, Guskara and joins Ajay near Mangalkot. The important left bank tributary is Hinglo which originates from a hillock named Deuli in Jharkhand having an elevation of 160 m. The Hinglo enters Birbhum district near Babuipur and flows south-east to join Ajay at Palash Danga. Kandar is

7.4 The Ajay

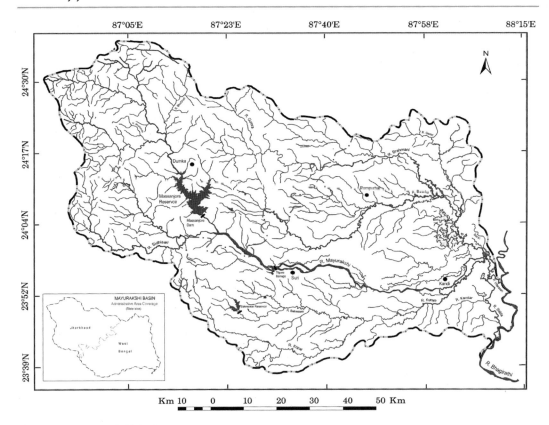

Fig. 7.3 The Mayurakshi basin

Table 7.3 Estimated flow in the Mayurakshi River

Basin	Basin area (km^2)	Estimated flow (MCM)											
		Jan	Feb	Mar	Apr	May	June	July	Aug	Sep	Oct	Nov	Dec
Mayurakshi	9345	195	150	4.2	−55	305	1162	2031	1843	1517	384	−121	−137

another left bank tributary which joins Ajay near its outfall at Katwa (Fig. 7.4).

The maximum runoff from the basin is generated in July when 1352 MCM of water flows through the river and a minimum discharge is expected in December (Table 7.4). The annual suspended load of Ajay was estimated to be 0.59 million ton. In fact, some dry weather roads are built across Ajay and transform it into stagnant pools of water. The lower Ajay basin is historically flood prone. The flood control embankments did not ensure total freedom from flood but recurrent breaches in embankment often caused distress to the people living in the floodplain. The devastating floods of the September 2000 opened many breaches in the old embankment and vast agricultural lands became non-productive due to deposition of coarse sand. The most devastating flood of the Ajay basin happened in September 1978 when the lower part of the basin area under Birbhum District was marooned. The shape of the basin is narrow and aligned in a west-east direction. The bed is shallow and choked with deposition of sand and silt. The bank-full capacity of the river near Katwa is about 856 cumec (30,242 cusec) and any volume in excess of this threshold limit causes flood.

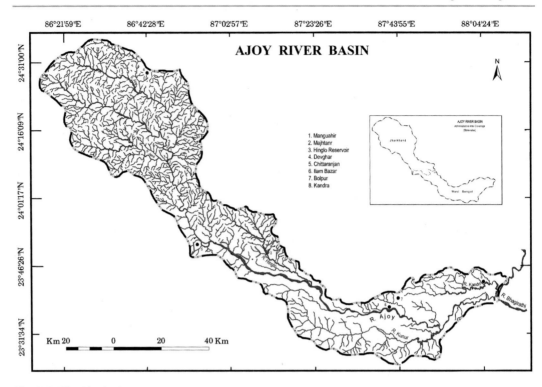

Fig. 7.4 The Ajay basin

Table 7.4 Estimated flow in the Ajay River

Basin	Basin area (km^2)	Estimated runoff leaving basin (MCM)											
		Jan	Feb	Mar	Apr	May	June	July	Aug	Sep	Oct	Nov	Dec
Ajay (Katwa)	6074	108	82	−16	−73	91	770	1352	1192	849	159	−88	−91

7.5 The Damodar

The Damodar is the most important tributary of the Bhagirathi–Hugli River. It originates from the Khamarpath Hill of Palamau in Chota Nagpur Plateau and flows for a length of 540 km, before discharging into Bhagirathi–Hugli through two major distributaries being bifurcated near Jamalpur (Bardhaman District). The eastern branch is known as the Amta Channel which joins the Hugli River at Gorchumbak near Uluberia. The other branch, known as Mundeswari receives Dwarkeswar, Silai as right bank tributary and finally discharges into the Hugli River through the Rupnarayan Estuary. A branch of the Kansai which takes off at Trimohoni (near Debra of Paschim Medinipur) also joins Silai further downstream at Ghatal. The total catchment area of Damodar is 20,874 km^2. About 16,910 km^2 of the catchment area belongs to Jharkhand and 6042 km^2 lies in West Bengal. The first major tributary to the Damodar is Barakar which joins at Dissergarh (Fig. 7.5 and Table 7.5).

The Damodar catchment generates substantial sediment load but only 0.35 million ton of suspended sediment load comes downstream of Durgapur barrage. The model study revealed that its highest discharge (5366 MCM) flows in July while minimum flow is observed in March. Damodar is historically flood prone. The peak discharge of the river during monsoons often

7.5 The Damodar

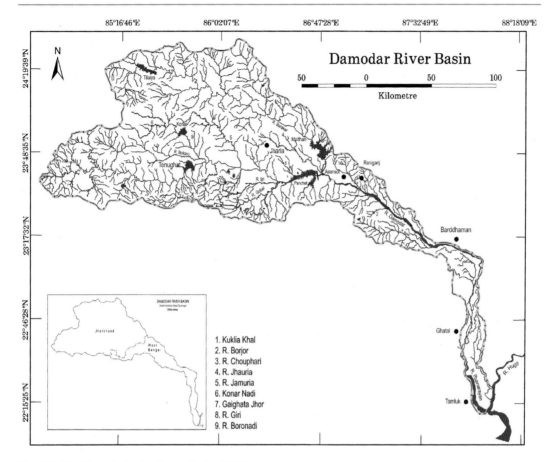

Fig. 7.5 The Damodar basin. *Source* Rudra (2012)

Table 7.5 Estimated flow in the Damodar River

Basin	Basin area (km²)	Estimated flow (MCM)											
		Jan	Feb	Mar	Apr	May	June	July	Aug	Sep	Oct	Nov	Dec
Damodar	20,874	68	57	−402	−716	−70	2530	5366	4945	3324	1036	−271	−312

overtops the banks causing devastating floods which tends to delink main communication lines connecting North India with Kolkata. The present channel of the river below Durgapur is so choked with sand and silt that flood is almost a recurrent event. The construction of linear embankments along both banks during the first half of the nineteenth century failed to control flood (Saha 2008). The British Government removed the right embankment to ensure the protection of the left embankment that protects the railway, national highway and Bardhaman Town from the flood. The railway connecting Haora and Bardhaman has created serious drainage congestion in the lower Damodar Valley (Saha 1933; Saha and Ray 1944).

The Damodar has changed its course in the lower reach appreciably. Earlier the river used to discharge its main flow through the Banka, the Gangur or Behula. These channels were active till sixteenth century as described in the mediaeval Bengali literature and maps drawn by European cartographers. Subsequently, the Damodar migrated southward and flowed through the present Amta Channel. In the course of avulsion, some intermediate distributaries became active and decayed (Fig. 7.6). In second half of the eighteenth century, the Damodar had flowed through the Amta Channel. Since the second half of the nineteenth century the Mundeswari Channel was opened up. In this process of southward migration, Damodar left behind a series of paleo-channels which are now stagnant pools of water and locally known as *Kana Nadi*. This change may be attributed to the natural avulsion, as well as human intervention (Rudra 2001; Bhattacharya 2002). The left embankment not only disconnected its eastern distributaries but also compelled the Damodar to migrate westwards (Rudra 2012). The old channels can still be the best possible outlets of flood waters, but the railways and highways have been extended across these channels creating severe drainage congestion in trans-Damodar area; hence, the flood problem remains unsolved even after the construction of the DVC reservoirs. The lower Damodar Valley is historically flood prone and agriculturally productive. The farmers had been coping with flood but the system decayed with the introduction of the colonial irrigation system (Biswas 1981, 1988). The history of human intervention in the fluvial regime of the Damodar since the colonial period is described in Chap. 10 and the problem of flood is explained in Chap. 9.

7.6 The Khari–Banka

The Banka is a beheaded distributary of the Damodar. It constituted one of the two major outlets of the Damodar during sixteenth and seventeenth century. This was described in the map of Van Leenen (1726). It discharged into the Bhagirathi at a place called Ambowa and a minor branch of it discharged into the same river near Nayasarai. These places were close to present Kalna (Fig. 7.6). The beheading of this river was initially a natural process causing southward migration of the Damodar through the Amta Channel. The fate of the river was further sealed by the construction of a flood control embankment along the left bank of Damodar in the early nineteenth century. This ceased the scope of dispersal of flood water which earlier spilled through the Banka and other east flowing channels (Fig. 7.6). The situation further aggravated when the railway line from Haora to Bardhaman was built during mid-nineteenth century. Such human intervention disconnected the Banka from its feeder and it started to function as an independent drainage basin. The Khari also seems to be an old spill channel of the Damodar which was also beheaded in the same manner. These two rivers combined now drain a catchment covering 2630 km^2 in the district of Bardhaman. This system is exclusively rainfed and the flow varies widely between the rainy and lean months. The peak discharge (475 MCM) flows during the month of the August while the river goes dry in the month of the December (Table 7.6). Other moribund distributaries of the Damodar are Behula, Gangur, Kunti, Ghia, Kananadi and Kanadamodar.

7.7 The Rupnarayan

The Rupnarayan is the common outlet of the Mundeswari, the Dwarkeswar, the Silai and one of the two branches of Kansai (also known as Kangshabati) and thus forms a large basin together. The Rupnarayan receives a portion of water from the Damodar through the Mundeswari. During the devastating flood of 1978, Mundeswari–Rupnarayan was the principal outlet of the freshet that came from the Damodar catchment. But the situation changed subsequently and during the flood of September 2000, the Amta Channel acted as the main outlet (Fig. 7.6).

The bulk of discharge draining Kansai catchment flows into the Rupnarayan Estuary while a negligible amount passes into the Haldi Channel. The combined basin area of Dwarkeswar and Silai is 8478 km^2. These two rivers are the major feeders to the Rupnarayan. Both the rivers are exclusively rainfed but the tidal intrusion reaches up to Bandor through the Rupnarayan Estuary. Since these two rivers having common outlets with the Mundeswari, the Khanakul, Arambag, Goghat and

7.7 The Rupnarayan

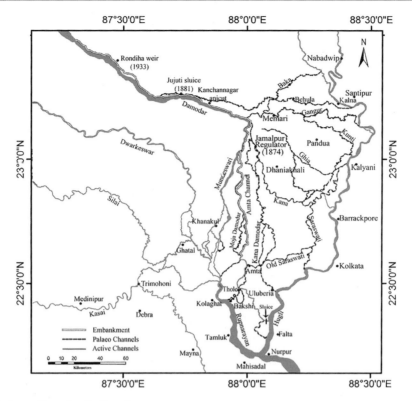

Fig. 7.6 The lower Damodar distributaries

Table 7.6 Estimated flow through the Khari–Banka

Basin	Basin area (km²)	Estimated flow (MCM)											
		Jan	Feb	Mar	Apr	May	June	July	Aug	Sep	Oct	Nov	Dec
Khari–Banka	2630	115	110	78	68	147	238	410	475	359	66	−32	−36

Table 7.7 Estimated runoff leaving the Silai and Dwarkeswar basins

Basin	Basin Area (km²)	Estimated runoff leaving basin (MCM)											
		Jan	Feb	Mar	Apr	May	June	July	Aug	Sep	Oct	Nov	Dec
Silai	1449	54	56	20	4	139	408	714	719	513	152	−31	−49
Dwarkeswar	4866	101	102	32	5	180	517	967	1011	713	189	−52	−68

Ghatal area suffer from severe drainage congestion during rainy season (Govt. of WB 2011). The dam-induced water from the DVC, synchronized with high rainfall and tidal intrusion caused many disastrous floods during the past. The mean monthly flows of the Darakeswar and the Silai are described in the Table 7.7.

7.8 The Kansai or Kangshabati

The Kansai having its origin in the uplands of Puruliya flow eastwards and ultimately discharges through two outlets, one into the Rupnarayan and other through the Haldi. The Kumari

Fig. 7.7 The Dwarkeswar–Silai basin

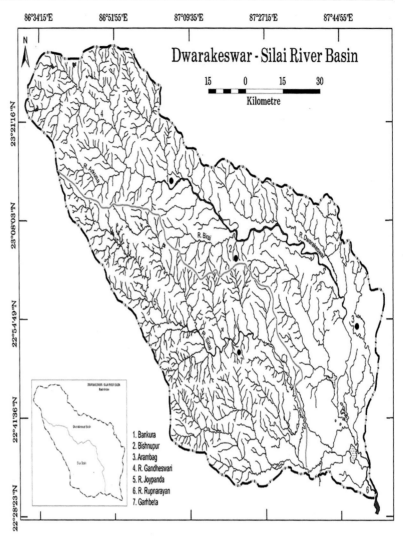

and the Kansai are combined at Mukutmanipur and flows eastwards through Paschim Medinipur. The other important tributaries are Bhairabbanki and Tarafeni. The river is bifurcated into two branches at Trimohoni near Debra of Paschim Medinipur. It is observed that about 60% of the annual flow goes to Silai from Trimohoni and the remaining volume passes into the Haldi. The Keleghai and the Kopaleshwari are two important tributaries to the Kansai–Haldi system. These two rivers have their sources in the upland of Paschim Medinipur and are notoriously flood prone in the lower reach. These rivers together have created a drainage congestion in Moyna basin. The Kansai irrigation project and other human interventions in the Kansai basin are described in Chap. 10 (Fig. 7.7). The Kansai is exclusively rainfed river the peak flow is generally recorded in the month of August (Table 7.8 and Fig. 7.8).

7.9 The Rasulpur

The Rasulpur and the Pichhabani are two minor tributaries which drain the coastal plains of Purba Medinipur and discharge into the Hugli River. These two rivers together constitute a catchment

7.9 The Rasulpur

Table 7.8 Estimated flow of the Kansai River

Basin	Basin Area (km^2)	Estimated runoff leaving basin (MCM)											
		Jan	Feb	Mar	Apr	May	June	July	Aug	Sep	Oct	Nov	Dec
Kansai	9527	183	200	89	41	417	1017	1867	1990	1455	496	−57	−127

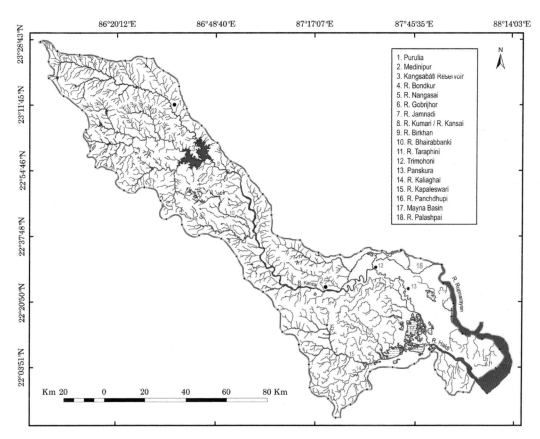

Fig. 7.8 The Kansai basin

area of 2074 km^2. Both the rivers are exclusively rainfed. But the tide invades regularly in their lower reach. Both of them spill off in the lower reach during high tide. The flow in these two rivers varies widely from the monsoon to lean season. The peak flows were observed in the month of September and in cases of both the rivers, the water levels diminish sharply during the post-monsoon seasons (Fig. 7.9).

The western tributaries to the Bhagirathi–Hugli River described in this chapter are decaying fast and that is the result of frequent human intervention in the fluvial regime. The extensive deforestation in the upper catchments, multiple cropping in the lower reaches especially in the floodplains, indiscriminate extraction of the groundwater leading to diminution of the base-flow towards the rivers during lean months' had been major causes of the decay. The flood control embankments reduced the spaces on which rivers tend to oscillate. In addition, constructions of many reservoirs to facilitate the storage of monsoon water reduced the ability of rivers to drain out the sediment load into the sea. The abstraction of water from the rivers as well as from the groundwater pool exceeding threshold

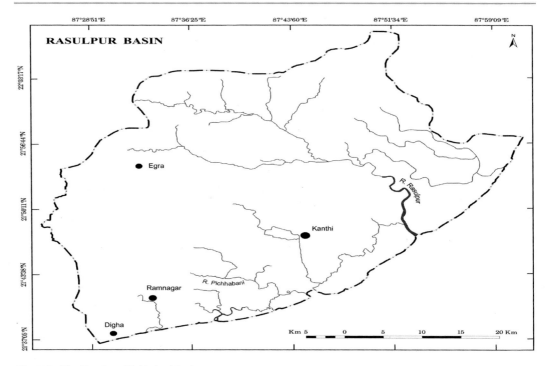

Fig. 7.9 The Rasulpur–Pichhabani basin

limit denied the importance of restoration of the ecological flow in rivers and this ultimately changed channels from perennial state to ephemeral.

References

Bagchi K, Mukherjee KN (1979) Diagnostic Survey of Rarh Bengal. Dept. of Geography, University of Calcutta, India

Bhattacharya K (2002) Damodar: Bandth Nirmaner Ekal O Sekal (in Bengali), (Damdar: Dam Building in Modern Times and Early Days), Pratiti; Edited by Rahul Ray. Chuchura, Hugli

Biswas A (1981) The Decay of Irrigation and Cropping in West Bengal (1850-1925). Cressida Trans 1(1):00001

Biswas, A. (1988) To live with flood: the case of West Bengal. Paper presented in the international symposium on river bank erosion, flood and population displacement. Jahangir Nagar University, Dhaka

Chapman GP, Rudra K (2007) Water as foe, water as friend: lessons from Bengal's millennium flood. J South Asian Development 2(1):19–49

Govt. of West Bengal (1989) River Systems of West Bengal

Govt. of West Bengal (2011) Unpublished base paper related to the activities of the newly constituted Flood Control Commission

Rennell J (1780) A Bengal Atlas. (Rudra K, 2016) (ed). Sahaitya Samsad, Kolkata

Rudra K (2001) Nimno Damodar Abobahikar Bonya O Nikashi Samasya: DVC-r safalya O Byarthota (Flood and drainage problem of the lower damodar basin: success and failure of DVC). Pratiti; Edited by Rahul Ray. Chuchura, Hugli

Rudra K (2008) Banglar Nadikatha (in Bengali). Sahitya Samsad, Kolkata

Rudra K (2012) Atlas of Changing River Courses in West Bengal. Sea Explorers' Institute, Kolkata

Saha M (2008) Rarh Banglar Duranto Nadi Damodar (in Bengali). Laser Art, Sreerampur, Hooghly

Saha MN (1933) Need for a Hydraulic Research Laboratory. Reprinted (1987) in Collected Works of Meghnad Saha. Orient Longman, Calcutta

Saha MN, Ray KS (1944) 'Planning for Damodar Valley'. Science and Culture, 10 (20) Reprinted (1987) in Collected Works of Meghnad Saha. Orient Longman, Calcutta

Van Leenen (1726) New map of kingdom of Bengal

The Sundarban

Abstract

The littoral tract of GBM delta is a unique area with thirteen major estuaries and many interlacing channels with intervening islands. This had been an area of dense mangrove forest which was partially reclaimed to facilitate agriculture and human habitation since the late eighteenth century. But the coastal Bengal had been the seat of an ancient civilization in the period 300 BC–1200 AD, but declined possibly due to gradual inundation. The tidal creeks were embanked to prevent ingress of saline water into the floodplain. The long-term effect of such premature reclamation was detrimental as sediment dispersal during high tide was impaired and the channels were choked. The coastline of the GBM delta has been changing fast; while the sea encroaching inland along Indian coast, the shape of the Meghna estuary in Bangladesh has changed due to accretion.

The littoral tract of the GBM delta is an area of complex drainage network and facilitates the largest mangrove ecosystem which has been declared as world heritage site and governed by India and Bangladesh. It extends from 21° 30′ N to 22° 30′ N and 88°10′ E to 90° E. Sir W. W Hunter (1875) noted in his 'A Statistical Account of Bengal', that '*the Sunderbans may be described as a tangled region of estuaries, rivers and water courses, enclosing a vast number of islands of various shapes and sizes*'. It spreads over 9630 km^2 in India and 9610 km^2 in Bangladesh including an impact zone of 3441 km^2. The Indian Sundarban is divided into reclaimed and non-reclaimed parts. The dense mangrove covers 4266 km^2. This region includes 1330 km^2 of Sundarban National Park, 362 km^2 of Sajnekhali Wildlife Sanctuary, 38 km^2 of Lothian Island Wildlife Sanctuary and 6 km^2 of Halliday Island Bird Sanctuary (Figs. 8.1 and 8.2).

The Sundarban in Bangladesh extends eastwards up to Baleswar estuary from the international border, and the area further east is completely deforested. The Sundarban in Bangladesh is classified into three separate zones. The forest area is bordered by an 'ecologically critical area' and also by an 'impact zone'. The littoral tract of the GBM delta is characterized by a complex network of interlacing channels with intervening islands. The creeks lying between the Hugli estuary in the west and Padma–Jamuna–Meghna estuary in the east are mostly beheaded having no upstream flow. Only a few channels like the Ichhamati in West Bengal, the Gorai–Madhumati and the Arial Khan in Bangladesh carry some freshwater during the monsoon season. According to official record the Indian Sundarban covers 104 islands, of which 54 have been deforested. But counting from recent satellite image reveals that the total number of

© Springer International Publishing AG, part of Springer Nature 2018
K. Rudra, *Rivers of the Ganga-Brahmaputra-Meghna Delta*,
Geography of the Physical Environment, https://doi.org/10.1007/978-3-319-76544-0_8

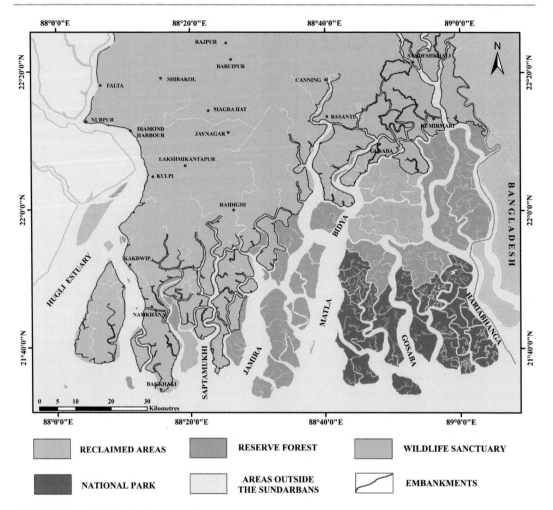

Fig. 8.1 Map of Indian Sundarban

islands is 128, of which 29 islands are settled by mankind. Since the earlier counting was done after independence, many smaller islands have been coalesced together because of decay of intervening channels and new islands have emerged by the process of accretion. The west to east width of the Indian Sundarban from Gangasagar to the Haribhanga estuary is 120 km and that in Bangladesh between Indo-Bangladesh border to Meghna estuary is about 230 km. The reclaimed Sundarban in West Bengal comprises six C. D. Blocks of North 24 Parganas and 13 C. D. Blocks of South 24 Parganas covering a total area of 4575 km^2; the water area of 795 km^2 includes intervening streams and creeks. These blocks together render homes to 4.5 million people. The Sundarban in Bangladesh is found in the southern part of Satkhira, Khulna and Bagerhat districts where more than 3 million people survive.

The origin of the name 'Sundarban' is much debatable. It might have been due to the abundance of the *sundari* trees (Heritiera fomes), and the area was named after the mangrove forests as Sundarban. It is the only tiger-infested littoral forest in the world. It might have its origin as *Samudra-ban* because of its seaside location (Pargiter 1934). It has also been termed as *Sudur-ban* due to its remote inaccessible location and *Sundarban* or beautiful forest. Some historians

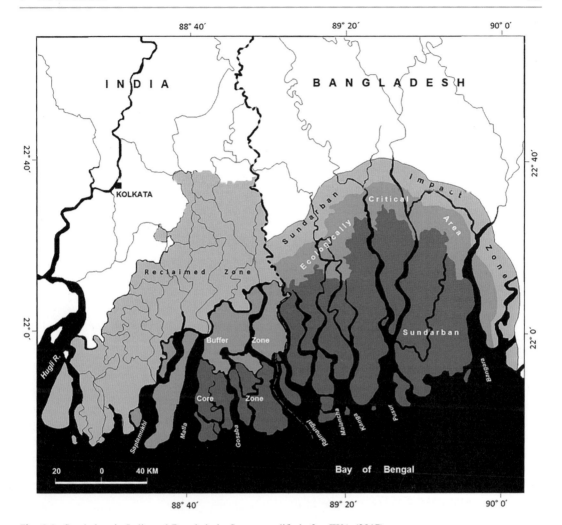

Fig. 8.2 Sundarban in India and Bangladesh. *Source* modified after IWA (2017)

have used the term *Bhati* for the coastal strip from the Hugli to the Meghna and that is still used by the local people. The name 'Sundarban' might also be derived from *Chandraban* because the shore area had different local names with the word *Chandra* (Chattopadhyaya 1999).

The geological research revealed that tidal swamps extended up to the Rajmahal hills and retreated southwards since the late Tertiary times. The growth of the delta was most striking during the Holocene period (Rudra 2016). Oldham (1870) first noted that the whole stretch of the Sundarban delta from the Hugli in the west and the Meghna in the east emerged from the sea due to deposition of sediment brought by the rivers of the GBM system. The origin and evolution of the vast littoral tracts are linked with plate movements, transgression and regression of the sea, tidal fluctuations, accretion and erosion of the coastal tracts of South Bengal. It has been concluded from the C^{14} and pollen analysis that the earliest mangrove forests were found to be 70 million years old and the species name was Spinizonocolpites. Then the Sundarban was extended up to Bardhaman, Bolpur, Jalangi region. The pollens of mangroves such as *Golpata, Keora, Garjan* were also found around Durgapur in Bardhaman district, and the sea had gradually receded southwards. The mangrove species have also been identified in the

Siwalik rocks of North Bengal. During the late Pleistocene ice age (1.10 million to 75,000 years B.P.,), the sea level is said to have receded about 120 m below that of the present. With subsequent marine transgression, peat formation started in the Sundarban region. The first peat layer below Kolkata is about 4.5–5 million years old, while the other below it is about 12–13 million years old. The relics of the swampy forests had also been found at Garden Reach, Fort William, Sealdah and at different excavation sites for the Metro Railway in Kolkata. These remains are reportedly 7000 years old (Banerjee 2008).

8.1 Hydro-Geomorphological Characteristics

The coastal tract of the Sundarban continues to grow by the deposition of silt mostly pushed back from the estuary by the tidal waves. The deposition in the littoral tract is a bio-tidal process. The fluvial sediments undergo metamorphism in estuarine environment. The organic matters present in seawater help to flocculate and settle down the suspended load. The shoals thus formed gradually emerge above low-tide level. The water in creeks during the high tides spills off the sediments on both shoals and adjoining the floodplain and thus facilitates colonization of the grasses and herbs. As the intertidal stretch achieves further height, woody mangroves occupy the area. The shoals and banks ultimately gain such a height that allows inundation only during storm surge when non-mangrove climax vegetations cover the area (IWA 2017). The rate of accretion may be as high as 12 cm/yr, as observed in Prentice Island (Paul 2002). The mangrove swamps are dynamic and differ horizontally and vertically due to the varying environmental conditions.

The Sundarban falls under the macro-meso-tidal zone and two existing classification schemes of littoral tract are described in Table 8.1.

The active delta is an area of diversified geomorphic features (Paul 2002). Most of the distributaries of the Ganga which ultimately discharge into the Bay of Bengal have been disconnected and do not receive any upstream flow. These channels are segmented into multi-directional creeks merging, separating and thus have created an interlacing drainage network. The eastward tilt of the delta and consequent deep incision by the Ganga–Padma has caused beheading of all distributaries at their respective off-take points. Only the Mathabhanga–Ichhamati, the Gorai–Madhumati and the Arial Khan maintain links with the feeder. The diminishing freshwater supply has impaired the delicate balance of the Sundarban ecosystem which sustained on the mixing of fresh- and saline-water regimes.

Thirteen major estuaries drain waters of the GBM delta into the Bay of Bengal and the Raimangal–Haribhanga flows along the Indo-Bangladesh border. Six estuaries flowing through Indian Sundarban are the Muriganga, the Saptamukhi, the Jamira, the Matla, the Gosaba, Haribhanga and those through Bangladesh are the Raimangal, the Malancha, the Kunga, the Haringhata, the Baleswar, the Tentulia and the Meghna. Only four of these thirteen estuaries carry freshwater to the Bay of Bengal even during non-monsoon season. Those are the Muriganga, the Baleswar, the Tentulia and the Meghna. The tidal creeks are funnel shaped being wide at sea-face and narrow at their northern limit. The estuaries are dominated by ebb and flow of oceanic tides and characterized by variable water levels, sediment content, salinity and electrical conductivity. The flora and fauna associated with the estuarine regions are unique due to the temporal and spatial variability of water quality. Some creeks may be fordable in low tide, but may have sufficient depth of water in high tide. The creeks are so dynamic that one can observe widening, deepening, bank erosion and decaying within a few years.

Whenever the water overtops the bank, there is a loss of its energy which leads to deposition of sediment and the process gradually forms natural levee along the bank. At the height of

8.1 Hydro-Geomorphological Characteristics

Table 8.1 Classification of tidal zones

Davies (1964)		Hayes (1979)	
Tidal range (m)	Class name	Tidal range (m)	Class name
< 2.0	Micro-tidal	< 1.0	Micro-tidal
2.0–4.0	Meso-tidal	1.0–2.0	Lower meso-tidal
>4.0	Macro-tidal	2.0–3.5	Upper meso-tidal
		3.5–5.5	Lower macro-tidal
		>5.5	Upper macro-tidal

Source Encyclopaedia of Coastal Science by Maurice L. Schwartz (2005)

tide, the water spills off to salt marsh or flats where it remains stagnant, depositing the finer clay. The water goes back to channel in the low tide. This lateral movement of water is repeated twice in twenty-four hours. The mangrove thicket helps in the trapping the sediment load from the overflowing water and building up a naturally raised surface along the banks. The meandering creeks tend to form linear depositional features due to lateral migration, termed as point bars. Lateral accretion of clay–silt and sand beds produces layers within the subsurface structure, dipping gently into the channel.

Mudflats are depositional features within intertidal zone. Fine clay and silt-deposited platforms formed by flocculation in estuarine and sheltered coastal environment produce mudflat marshes which are reshaped by the wave and tidal actions. The intertidal zones are submerged and exposed twice daily and may be colonized by the mangroves resulting in holes, moulds, pits and pillar structures. Phytoplankton and zooplankton are abundant there. The crabs and other burrowing animals creep over the flats at high tide. Because there is little or no oxygen, they breathe through tubes to get oxygen from the surface. Birds and some predatory animals come on mudflats at specific times for their catch. The tidally swept mudflats are often scoured forming depressions where water accumulates. The rapid evaporation causes salt encrustation in these pans. They are mostly devoid of vegetation, but some salt-tolerant vegetation may grow in patches.

The higher swamps are featureless plains which stand above mean high tide level but drained by channels and creeks. As the surface level rises, the frequency of inundation and sediment accretion rates decreases. The alternate flooding and scouring produce micro-geomorphic features like pits, holes, depressions, puddles. The accumulation of organic debris and litter improves the faunal habitat in this region. The lower swamps are shallow bodies of water in a low-lying area, which are poorly drained and hold water due to their fine texture. They are mostly situated at the edges of islands and river-side bars where marsh and swamp vegetations enhance the replenishment of silt layers.

The coalescence of islands and their extension largely depend on the rapid growth of the younger mangroves in lower swamp region. The Sundarban consists of low-flat alluvial plains, non-reclaimed virgin forests and swamps. The creeks, which do not get upstream flow, are still active by the tidal inflow. The great chain of brackish marshes, stretching from the west of Kolkata to Barisal and Jessore in Bangladesh, marks the dividing belt between mature and active deltaic region. The numerous creeks and channels have formed several islands at coast of Bengal. These are Ghoramara, Khasimara, Lohachara, Sagardwip, Jambudwip, Namkhana, Frazerganj, Bakkhali, Henry Island, Fredrick Island, Lothian, Bullcherry, Halliday, Bangaduni, Jharkhali, Mayadwip, Sajnekhali, Dutta, Marichjhapi, Champta, etc. The eastern part of Sundarban is still covered with dense impenetrable forests where land is being elevated due to the uninterrupted sediment deposition. The most striking change has been noticed in the Meghna estuary which is the main conduit silt-laden water. The embankments in Indian Sundarban were built to prevent the spill of the silt-laden water during high tide to facilitate human habitation and agriculture on the floodplain, but this programme ultimately impaired sediment dispersal away from the channels. As on today, the mangrove area where natural hydrological process operated uninterruptedly stands at least at 2 m above reclaimed area. There are 75 islands covered by dense mangroves within administrative control of the West Bengal Government. The official record claims that fifty-four islands were cleared to facilitate human settlements but the number has been subsequently reduced to twenty-nine. This happened due to decay of intervening channels which earlier cut across the islands. The sea erodes the southern front of the coast with wave attack and carries sediment load inland. The period from 1917 to 2016 has witnessed alarming deposition in the inland creeks. The famous Sagar Island was a cluster of six islands of different sizes when Rennell (1780) surveyed Bengal in the second half of the eighteenth century. But the intervening channels decayed subsequently and islands were coalesced to form a single unit. Further, the northern part was detached to constitute a new island now called Ghoramara (Rudra 2012). Two smaller islands called Lohachara and Suparibhanga have disappeared recently. But a new island called Nayachar covering 49 km^2 has emerged just north of Ghoramara (Fig. 8.3 and 8.4). In a nutshell, sea-level rise, tidal fluctuation, sediment movement and subsidence of land are the major factors governing the changing geographical scenario of the Coastal Bengal.

8.2 The River System

Many channels flowing through the Sundarban are not rivers in strict sense of fluvial geomorphology. A river is supposed to have a source, outfall, catchment area, many tributaries and distributaries in its lower reach. Unlike the general characters of the rivers, the creeks receive water supply from the downstream that is from sea or ocean. Larger rivers may be divided into tidal and non-tidal regime. Water flowing in creeks is induced from the sea, and the flow is governed by the tide-velocity asymmetry and thus has a two-way flow. Most channels in Sundarban can be broadly identified as tidal creeks with few exceptions like Ichhamati–Haribhanga, Gorai–Madhumati and Arial Khan which receive feeble supply of freshwater during peak of the monsoon.

It is important to note about 15% of the estimated one billion tonnes reaching the Bengal

Table 8.2 Cyclonic activity over Sundarban (1901–2012)

	Depressions	Storms	Severe storms	Total
January	0	0	0	0
February	0	0	0	0
March	0	0	0	0
April	0	0	0	0
May	4	8	15	27
June	76	6	8	90
July	73	30	18	121
August	120	8	4	132
September	67	4	12	83
October	19	8	6	33
November	8	4	6	18
December	0	0	8	8
Total	367	68	77	512

Source Indian Meteorological Department

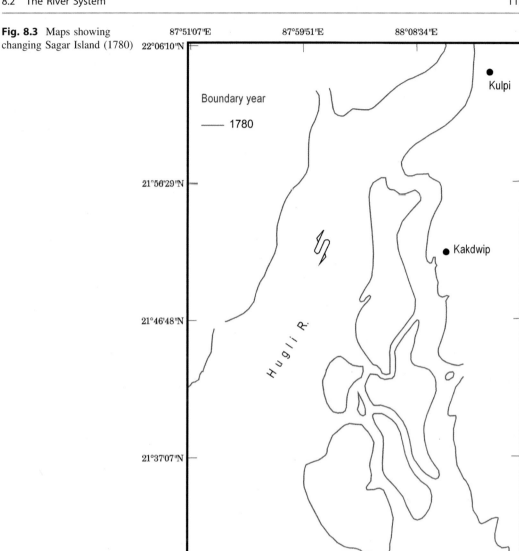

Fig. 8.3 Maps showing changing Sagar Island (1780)

basin is sequestered annually and does not reach the sea (Goodbred Jr. and Kuehl 1998). Since a large part of the Indian Sundarban is 1.5–3 m above mean sea level, the water swells 5–6 m when spring tides are coupled with cyclonic storms. The water tends to spill off the banks and recede at the time of low tide. This happens mostly in the August or September. The silt-laden water travels up to the northern tidal limit of the creeks. Thus, a process of accretion operates to build up the floodplain. The deposition of silt gradually blocks the flow of the river and the old creek splits up around obstructions, thus forming numerous channels, tidal creeks and distributaries. The cross-channel (locally called Duani) connecting two larger creeks faces faster decay due to head-on collision of the tidal waters leading to settling down of sediment load. During pre-monsoon season, the flood tide invades with greater vigour which causes soil erosion and

Fig. 8.4 Maps showing changing Sagar Island (2010)

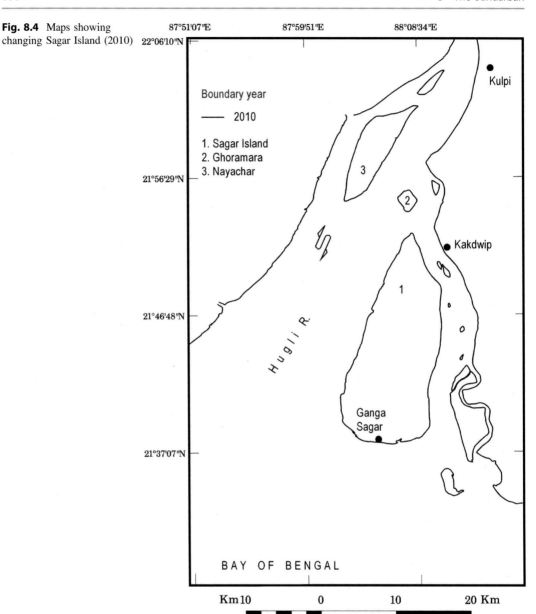

loss of land mass. Many creeks have been filled up during the last two centuries. The scenario in Sundarban of Bangladesh where the tidal range is less than 4 m is slightly different. The polders were created since the 1960s, sediment load from the creeks is flushed off by the monsoon freshet and the rate of sedimentation in channels is slower than observed in Indian Sundarban.

8.3 The Major Channels of the Sundarban

(a) *Hugli*: The Hugli River is bifurcated at the northern limit of Sagar and Ghoramara islands and the eastern branch flows south-eastwards along the banks of Kakdwip, Namkhana,

Mousumi, Bakkhali islands. This branch is known as Muriganga. The other branch flows between Sagar Island in the east and the coast of Purba Medinipur in the west and is known as Gabtala River. These two branches together form the funnel-shaped Hugli estuary.

(b) *Saptamukhi*: This was the older outlet of the Adi Ganga. It originates near Sultanpur and offers outlet to the Thikara *khal*, the Banstala *khal*, the Ghugudanga *khal* and is connected with Muriganga through Hatania-Doania creek.

(c) *Thakuran*: It originates near Jaynagar, and Mathurapur flows southwards for a length of 80 km. At its northern end, it connects the Matla and has links with Saptamukhi. The Jagadal *gang* flows out of the west bank of the main channel. The Thakuran takes the name of Jamira before reaching the sea.

(d) *Matla*: This second largest estuary of the Indian Sundarban is formed by the combined flow of Karati, Rampura *khal*, Atharbanki and Bidyadhari. It flows for a length of 120 km from Canning to the Bay of Bengal. The Matla provided better navigational depth than the Hugli estuary in the mid-nineteenth century, and Port Canning was built as a substitute to Kolkata port. But the port was abandoned within a decade due to rapid sedimentation.

(e) *Piyali*: It took off from the Bidyadhari River near Pratap Nagar and Bamanghata, flowed 45 km to join the river Matla. But this channel has totally decayed.

(f) *Bidya*: The Huta Khal and Durgamangal creek are combined to form the Bidya. It receives Pathankhali creek, Radhanagar creek, Dundul *khal*, Kayer creek, Mocumberiya creek, Moukhali *nala* and Kartal ganj from the right and Melmel *khal*, Gumdi, Durga Duania, Nabanki and Netidhopani creeks from the left. It flows 55 km through Gosaba and Basanti blocks and meets Matla River near Jharkhali.

(g) *Gosaba*: The Gosaba in the west and Haribhanga in the east are two branches of the Jheela. Numerous creeks from Raimangal and Matla join this river. It flows more than 30 km through the Sundarban and discharges into the Bay of Bengal.

(h) *Bidyadhari*: This moribund river today serves as an outlet for sewage and storm water from Kolkata and North 24 Parganas. It flows south-east and receives Noai and Suti before it joins Kulti *Gang*.

(i) *Kalindi*: The Ichhamati is bifurcated into two separate braches near Hingalganj; the western branch flows as Sahebkhali and the eastern channel as the Kalindi.

(j) *Raimangal–Haribhanga:* The Ichhamati ultimately discharges through Kalindi–Haribhanga. It is connected with Barakalagachi and Terobanki. Haribhanga and Raimangal are combined together before meeting the Bay of Bengal. It also demarcates the Indo-Bangladesh border.

(k) The Malancha, the Kunga and Haringhata, Bhairab, Kapataksha and Arial Khan–Baleswar flow through the Sundarban in Bangladesh. While the first two do not have any link with the Ganga–Padma, the Haringhata offers outlet to the Gorai–Madhumati River. The Arial Khan, which had been the old outlet of the Padma, flows through the Baleswar estuary.

8.4 Early Civilization

It is believed that the coastline of the western GBM delta achieved the present position 7000 years B.P., though accretion and erosion is an unabated process. Notably, the coastal region was not fit for human habitation at that time. The high tide used to submerge the low-lying tract and cyclical submergence and emergence was hydro-geomorphological processes operating in the littoral tract. The historians and archaeologists referred to the literary and archaeological evidences and identified the Sundarban as a seat of a very ancient civilization which flourished during the period 300 B.C.–1200 A.D. (Ray 1949). Archaeological evidences of early human settlements were observed in Deulpota, Harinarayanpur, Chandraketugarh, Netidhopani and many other places. The classical texts like Upanishads, Puranas, Ramayan (*Balkhanda*) and Mahabharata referred to the littoral tract or present Sundarban. The great epic 'Mahabharata' talked about the holy shrine of Gangasagar in its *tirthayatra* section. Greek and Chinese travellers have repeatedly given accounts of a prosperous kingdom of Gangaridai in lower Bengal. Sundarban was first quoted in 'Indica', a report by Megasthenes (381–312 B.C.) who visited India, during the reign of Emperor Chandragupta Maurya. In 75 A.D., a Greek anonymous sailor in his *'Periplus of the Erythraean Sea*, a handbook of the Indian Ocean' (later translated into English by W.H. Shoof in 1912) accounted for the flourishing port and market city called 'Gange' that used to facilitate the trade of betel, pearl, Gangetic spikenard, conch, corals and the finest muslin of the world. The medium of exchange was gold. Other classical historians and writers who described Sundarban were the Roman poet Virgil (70–19 B.C.), Deodorus (first century A.D.), Pliny (23–79 A.D.), Velerious Flaccas, Kartius Rufas (Halder 2000). Ptolemy in 150 A.D. first attempted a cartographic representation of deltaic Bengal including locations such as Tamalitis (Tamralipta), Tilograman, Poloura and Gange. The credit for preparing early maps of Bengal also goes to Portuguese, French and Dutch cartographers like Mattheus van den Broucke and Van Leenen (1726), d'Anville (1752), Gastaldi (1561), and Jao de Barros (1660). Those maps were not cartographically correct in the modern sense but gave enough information for historical and geographical research. The first systematic survey was done by James Rennell during 1764–77, and A Bengal Atlas was published in London in 1780.

After the Mauryan and Gupta empire in the first and second century A.D., the lower Bengal was ruled by King Sasanka, the Pal dynasty and Sen dynasty in 1100 A.D. During the Gupta period the area was the part of *Samatata,* while in the Pala period it was known as *Baghratatimondala,* and during the Sen rule it was referred as *Kharimondala*. During the reign of one of the local chiefs (Baro Bhuiyas), King Pratapaditya ordered shifting of the capital of South Bengal from Dhumghat to Jessore and intended to move further south, when a large portion of the forest had been encroached upon. He also constructed a naval port for boat troops and forest check posts at the heart of Sundarban. Relics of those isolated brick structures, temples, tanks and wells can be seen even today. The Muslim rule started in South Bengal when Ghyasuddin Tughluk invaded in 1328. It continued till the advent of the Mughals in 1576. It was renamed as *Pundravardhanbhukti* during the reign of Hussain Shah and was known as *Bhatidesh* under the Mughal reign. In the last few years of Shahjahan's rule, persistent Portuguese attacks on Sundarban compelled many residents to migrate from this area which became more densely forested. From the early sixteenth century, European traders, travellers and pirates have frequented the prosperous southern Bengal. In 1559, Raja Poromananda Roy of Bakla granted the Portuguese full rights over Sundarban and also for part of Khulna district. The other Europeans who visited Sundarban were of Dutch (1632), French (1673), Danish (1676) and British origin (1690).

The early civilization that flourished in Sundarban had declined after the twentieth century. The region has been affected by sea-level fluctuation, changing river course, subsidence, frequent storms, flooding and other natural disasters along with invasions of pirates and plunderers exploiting the coastal inhabitants from time to time. The Sundarban was depopulated, and subsequent encroachment on forested tracts occurred in stages. In 1688, about two lakh of people were swept away by flooding in Sagar Island (WWF 2010). Gradual subsidence of some areas due to autocompaction of sedimentary layers might have caused people to migrate elsewhere, but a massive subsidence in 1707 displaced a large number of inhabitants at one go (ibid). In 1737, an earthquake caused the rise of the storm surge by about 12 m and most of the area remained submerged for a long period. With the change of river courses, the older ports lost their prominence and human assemblages for trade and commerce to those areas declined.

8.5 Premature Reclamation

After the battle of Plassey and defeat of Nawab Sirajuddaullah in 1757, Mir Jafar conferred the *zamindari* of 24 Parganas to the British East India Company. The natural resources, social and economic potentials lured them to reclaim these lands for cultivation, settlement and collection of revenue. Mr. T. Henckell, Judge and Magistrate of Jessore, submitted a proposal of reclaiming parts of Sundarban on the 20 December, 1783, and got the approval of Warren Hastings, the then Governor General on the 3 April, 1784. Since then, an extensive forested area was cleared or reclaimed for cultivation and human settlements. The Collector General Mr. Claude Russell granted lease to individuals to reclaim the land for cultivation and supply of timber. The area came under the authority of the colonial administration and the boundary of Sundarban was defined. The Regulation III of 1828 recorded the uninhabited tracts or *Bada* jungles as the Sundarban forests and declared the whole of Sundarban to be the sole property of the Government. Thus, the East India Company attained supreme authority to clear, cultivate and lease any part of the lands within its jurisdiction. The reclamation in the Sundarban occurred in phases.

When the first phase of reclamation started, the population around this region then was around three lakh. In 1828–33, the then commissioner Mr. William Dampier and surveyor Lt. Alexander Hodges drew the border subsequently known as the Dampier-Hodges Line to fix the northern limit of the Sundarban. The first all-inclusive set of rules for lease of land was issued in 1853. A stable income from forest products and secure land was assured to these people who were brought to Sundarban from Medinipur and Chotanagpur area. The administration of Sundarban was governed by the Commissioner from 1816, with the duties, powers and authority of a collector of land revenue. But this was discontinued in 1905. The Sundarban Act (Bengal Act 1 of 1905) was passed, and the Collectors of the three districts—24 Parganas, Khulna and Bakherganj—were entrusted with the authority of the Sundarban. The reclamation initially started in the western Sundarban and gradually extended eastwards. Between 1873 and 1911, 2608 km^2 of forested area was cleared to facilitate cultivation in Bakherganj and adjoining areas (IUCN 2014). The mangrove area has reduced appreciably since late eighteenth century. While forest area in Indian Sundarban reduced from 8833 to 3094 km^2, mangrove cover in Bangladesh shrunk from 11,256 to 5487 km^2 between 1780 and 2010 (IWA 2017) (Fig. 8.5).

The 'Dampier-Hodges Line' which delineated the boundary of the Indian Sundarban runs from Kulpi to Basirhat towards the northeast and incidentally corresponds with the Kakdwip–Basirhat–Dhaka lineament (Pargiter 1934[2002], WWF 2011). This area was described as the British '*Khas Mahal*' in the early nineteenth century. The *zamindars* of this region unheeding the British laws initiated reclamation of the forested lands, lying close to their estates, and did not pay revenue to the East India Company. Fearing the *zamindari* supremacy over the land and revenue loss, the Company started to identify the non-leased forests of Sundarban as British estates. This was done to establish their exclusive authority (Charkraborty 2005). The motive of reclamation was to exploit the vast natural resources of Sundarban. To create better navigation link with Kolkata, they excavated a 27-km-long link canal from *Adi* Ganga at Garia to Bidyadhari at Samukpota during 1775–77 which came to be known as the Tolly's *Nala*.

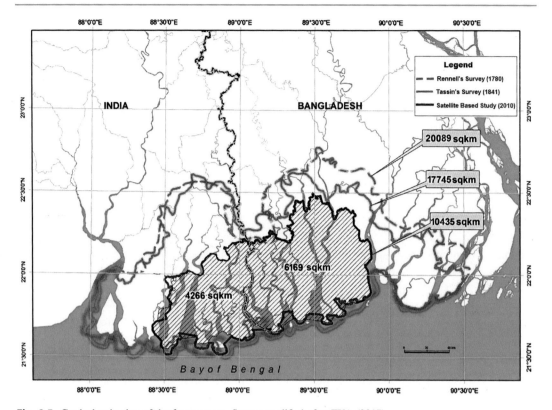

Fig. 8.5 Gradual reduction of the forest cover. *Source* modified after IWA (2017)

8.6 Protection with Embankment: Myth and Reality

The reclaimed areas were not fit for cultivation due to tidal submergence with saline water. The landlords or *zamindar* constructed some embankments to prevent the ingress of saline water into the floodplain. The practice got a momentum during the regime of East India Company, and the ingress of saline water on floodplains was controlled in western or Indian Sundarban. This ultimately accelerated deposition of sediment load on the riverbeds (Fig. 8.6). The embankment also intercepted the rainwater falling on the floodplain and invited problem of water logging.

Notably, the reclamation of land in Sundarban started even before the land building was complete.

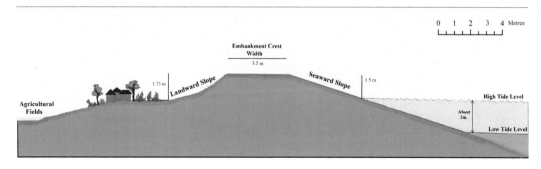

Fig. 8.6 Sections along Rangabelia Ghat (Gosaba)

With the increasing population, huge forested tracts mostly in the western side were cleared off. Many of the unique plant and animal species were endangered, and shrinkage of the forest cover compelled the animals to migrate elsewhere or enter into the inhabited areas causing man and nature conflict. The embankments have been constructed for prevention of river bank erosion, inundation of floodplain, protection of hinterlands, particularly the agricultural land and inhabited areas immediately above intertidal zone. But guarding and maintenance of these embankments were found difficult. Since the 1950s, maintenance of about 3500 km embankment was given under the authority of the West Bengal Government. These structures were constructed of bricks, lateritic rock, concrete and mud, but most of these have been frequently breached or slumped down as they could not withstand the salinity of the tidal waters. Some were unscientifically constructed and could not stand against the impinging wave for long. The non-concretized sea facing embankments generated cracks due to the prolonged exposure to wave attack and allowed the water to penetrate the wall. Biological weathering by mangrove roots and liquefaction of the subsoil were other causes of embankment failure. It often led to tilting of embankments, cracking of mud below the embankments and lateral spreading of subsoil towards the sloping sides (Bhattacharya 2000). Another cause for the instability of these embankments has been the inherent tendency to subside with time. The river-side slope of retired embankment was often too steep resulting in the slide of the materials even before proper consolidation. It is seen that embankment failure occurs at the concave banks due to continuous under-scouring creating a hollow at the base of the wall. The circuit embankments constructed by the Company failed to take note of the sediment flow in the streams, and hence, construction of those structures prevented the overspilling of the silt and that in turn was deposited in the channel beds. The embankment further caused the tidal inflows to transform into tidal bores causing embankment collapse when they struck with great force (Chakraborty 2005).

It must also be noted that construction of embankments has prominent ill effects on the hydrology of the Sundarban. It leads to decay of rivers, inlets and creeks due to excessive siltation. The embankment did not ensure freedom from storm surge. The cyclones or tidal surges frequently caused the increased volume of water to spill over the embankment. But when the water levels in the channel receded, spilled water was unable to flow back due to lack of proper drainage.

Embankments reduce the channel width, and sedimentation reduces its depth; hence, normal tidal waves appear to enter the creeks as tidal bores. The migration of creeks and reduction in depth exert pressure laterally on the banks causing embankments to breach and collapse. These structures also truncate the growth of vegetation and hinder the landward expansion of the mangrove forests. Breaching of embankments causes the gushing in of saline waters into the agricultural lands destroying property and livelihood, and eventually reduces the crop yield, forms salt encrustations and damages the productivity of the soil.

Bangladesh started to built embankment and create polders since the 1960s with the same intention of protecting villages and agricultural land from tidal submergences. The impacts are not yet very pronounced because the rivers unlike those in West Bengal have the ability to carry the sediment load down the estuary during the monsoon. Secondly, the polders are much younger compared to West Bengal and it will take a longer time to appreciate decay of channels.

The creeks or rivers in Sundarban need wider spill area to accommodate excess water during high tide. The growth of mangroves in intertidal zone may protect embankment. The destructive waves during Aila (25 May, 2009) not only breached but also crossed over the embankments at many places. The drainage of saline water from the land took a long time as breaches in the embankment allowed regular ingress of saline water. Evaporation of water left behind salt on the topsoil making it non-productive.

The Sundarban suffers from freshwater crisis for the major parts of the year in spite of its location on seafront and copious rainfall. The farmers cultivate the land only during the monsoon rain leaving the land fallow for the remaining months of the year. The Sundarban receives more than 1800 mm of rainfall annually, but superficial clay layer restricts infiltration to less than 17%. The WWF (2011) states the groundwater reserve in the southern Sundarban is around 68 m.c.m./year and potable water available at depths ranging between 160 and 400 m. The two other overlying aquifers at the depth of 60 m and 70–60 m below ground level are brackish. An intervening clay layer with differential thickness between 4 and 120 m acts as the barrier between fresh- and saline-water WWF (2010).

8.7 Impacts of Tropical Cyclones

The coastal tracts are frequently visited by the tropical cyclones and the storm surges, and climate change may be one of the reasons for this changing scenario. The millions of poor who have contributed little to climate change are worst sufferers. The increasing number of the tropical cyclones experienced in recent years is generally ascribed to rise in sea surface temperature (CSE 2009). It is expected that the coping mechanism and management strategies should emerge out of our experiences and lessons from the past. But the disaster management plan is generally reactive and starts to operate when a cyclone strikes. It largely relies on the colonial legacy of embankment building to protect the villages from tidal upsurge. But the past experiences of Sundarban teach a different science.

Sundarban has experienced many disastrous cyclones and storm surges (Table 8.3). It is revealed from the official records of Indian Meteorological Department 367 depressions, 68 storms and 77 cyclones developed on the Bay of Bengal during the period 1901–2012 (Table 8.2). The most disastrous cyclone struck Indian Sundarban on the 25 May, 2009, causing extensive damage of the embankments. Since then a popular demand of strengthening and raising the embankment with brick and cement emerged. But such an opinion was not based on a scientific study. The load of overlying embankment often leads to subsidence and consequent collapse.

The tropical cyclones and storm surges ravaged the Sundarban many times during the preceding decades. The preparedness to cope with the disaster in the remote islands had been inadequate to save the afflicted people. A paradigm shift from ongoing relief-centric approach to proactive

Table 8.3 Some disastrous cyclones over the Bay of Bengal

Name & date	Maximum speed at landfall (km/hour)	Location of landfall	Impacts
2B, May 2002	200	Bangladesh and Myanmar border	285 deaths
Typhoon Muifa, November 2004	130	Philippines	68 deaths, 160 injured, 60 unaccounted for
Xangsane, September 2006	160	Philippines, also affected Vietnam and Thailand	279 deaths
Mala, April 2006	185	Myanmar	22 deaths, 6000 houses destroyed
Sidr, November 2007	250	Bangladesh	More than 5000 deaths
Nargis, May 2008	215	Myanmar	1,46,000 deaths
Bijli, April 2009	70	Bangladesh	7 deaths, 84 injured, 107 houses destroyed
Aila, May 2009	110	West Bengal Coast, Bangladesh	More than 300 killed, innumerable missing

preparedness is urgently required. Though embankment does not offer total protection against tidal upsurge, breaching and overtopping caused enormous distress of people after Aila, and some ad hoc measures were necessary to protect the affected people (Rudra 2011). These were plugging of breaches, construction of retired embankment and rehabilitation of the erosion victims. But permanent solution lies in realigning embankments beyond intertidal space. In fact, rivers demand more space to spill off during the high tide or the storm surge, otherwise underwater scouring cannot be prevented. It is learnt from the past experiences that the plantation of mangroves on intertidal space protects the embankment from the erosion.

8.8 Recent Challenges

About 7.50 million people living in Sundarban (4.5 million in West Bengal and 3 million in Bangladesh) continuously struggle for their existence. In a recently published report of Centre for Science and Environment (2012), two environmental changes were reported: (a) The sea surface temperature (SST) in the Bay of Bengal is going up at an alarming rate of 0.5 °C per decade, in contrast to the global average rise of 0.06 °C per decade; (b) the sea level in the Bay of Bengal is also rising a rate almost double than the global average. The effect is accelerated due to slow subsidence of the coastal tract. Further the average annual rainfall has increased. August, July and September show even higher increasing trend. The intensity of the cyclonic storm over the Sundarban has also increased appreciably.

8.9 Rising Temperatures

The temperatures in Sundarban have been rising, and this has affected the ecosystem. The major causes are anthropogenic that includes unplanned and premature reclamation, widespread deforestation and burning of fossil fuels. The increasing sea surface temperature has been leading to increasing frequency of cyclones, tidal surges and sea-level change. The biological adaptations of the flora and fauna in this region are disrupted by the changing climatic conditions. It is estimated globally that with the current rate of emissions of greenhouse gases, the sea level may rise 15–90 cm higher by 2100 AD, threatening a large number of coastal population (Sanyal 1999). In accordance with the rising temperatures graph, the monsoon rainfall tends to be skewed. Taking into account the meteorological data of the period 1901–2002, it was found that the summer temperature has increased slightly and the average winter temperature has risen by nearly 1 °C. The June rainfall has declined by about 48 mm, and the September rainfall has increased appreciably. The ongoing crop calendar has to be reoriented with the changing rainfall patterns. The prolonged summers, contracted winters and erratic breaks in monsoonal precipitation have affected agriculture and made the crops prone to diseases. The increased dependence on groundwater for irrigation and more use of pesticides have escalated the cost of production.

8.10 Sea-Level Changes

The 260-km-long west–east aligned Sundarban coast between Gangasagar and Baleswar estuary has recorded variable relative mean sea-level (RMSL) changes at different location. IPCC in its fifth assessment report observed mean sea-level rise at 1.7 mm/year during the period 1901–2010 and that increased to 3.2 mm/year between 1993 and 2010. While average tidal range varies from 4.38 m at Sagar to 3.97 m at Mongla, the RMSL trends without adjusting ground level rise were found to be -2.98 ± 0.93 mm at Sagar and 6.25 mm at Mongla. But the negative RMSL at Sagar is not corroborated by ground reality as the sea has been encroaching inland (Fig. 8.7).

This is one of the major threats to the Sundarban ecosystem as the water level in the Bay of Bengal is rising at a much faster rate than in other regions. This is aggravated by slow subsidence of land. The experts have mentioned different rates of subsidence. Brown and Nicholls (2015)

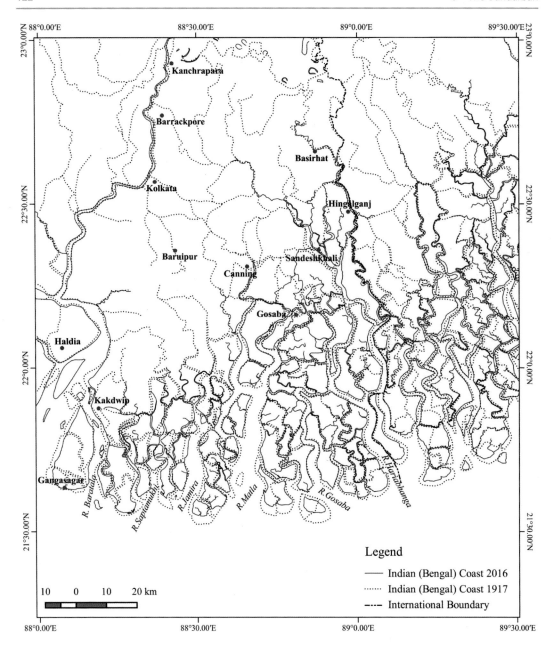

Fig. 8.7 Changing coastline of the western GBM delta

have reviewed 24 studies published till 2014 and noted that reported mean rate of subsidence was 5.6 mm/yr and overall median rate was 2.9 mm/yr. The scholars have different opinions regarding sea-level rise. Hazra (2010) stated that the change in sea level at Indian Sundarban was +12 mm/yr during 2002–09 compared to +3.14 mm/yr in the preceding decade, though the global average was observed < 2 mm. Nandy and Bandyopadhyay (2012) observed the trend in the Hugli estuary at four observatories of Permanent Service for Mean Sea Level Changes

(PSMSL) and computed the changes to be −3.82, ± 0.89, + 2.43, + 4.45 mm/yr, respectively. A conservative estimate stated the rate between 1.06 and 1.75 mm/yr (Unnikrishnan and Shankar 2007). The rise of sea level in coastal Bengal is always coupled with land subsidence. It is estimated that 50 cm rise of the sea level may cause landward advancement of the sea by 20–25 km (Matin 2016).

The century-scale change of the Bengal coast is detected by comparison of a map published in 1917 and a Landsat image of 2016. This conclusively proves that the Bay of Bengal has advanced inland appreciably in Indian Sundarban, but the Meghna estuary in Bangladesh has recorded significant southward advancement by the process of accretion (Fig. 8.7 and 8.8). The Indian Sundarban has gained 220 km^2 area due to accretion and has lost 430 km^2 area by erosion during 1917–2016. WWF (2010) India identified 12 islands on the Indian side which are subject to rapid erosion. The estimated rate of coastal erosion in the Indian Sundarban was about 5.5 km^2/year during 2001–2009, and maximum change was detected in the south-western part (Hazra 2010). The erosion of the Lohachara, Ghoramara and Suparibhanga islands was aggravated due to the high sand content in the subsurface. Mousuni island too has been eroded after breaking off the seven metres high embankments built by the Britishers in the 1920s. The local people are of opinion that the raising of the embankment was of no use as the tidal surges overtopped or breached it frequently (WWF 2010) (Fig. 8.8).

On the contrary, the Meghna estuary has prograded significantly during last 100 years (Fig. 8.8). Brammer (2014) compared the map of 1943 with that of 2013 to detect the change. He further worked on the Landsat images of 1984 and 2007 and revealed that there was net gain of 451 km^2 within that period. Thus, there was average accretion of 19.6 km^2/yr. Sarker et al. (2013) opined that the Meghna is prograding at 17 km^2/year. It is revealed from the superimposition of the coastline of 2016 on that of 1917, the Meghna estuary has changed dramatically. The most striking change of the coastline is recoded along Lakshmipur, Noakhali and Sitakund where the promontory has grown about 35 km towards the sea. Paradoxically, the island of Sandwip has been reduced in size by about 40% during the period 1943–2013.

The landscape in coastal Bengal is fast changing (Bandyopadhyay and Bandyopadhyay 1996). While the coastline in Indian Sundarban is encroaching inland, that along Bangladesh is

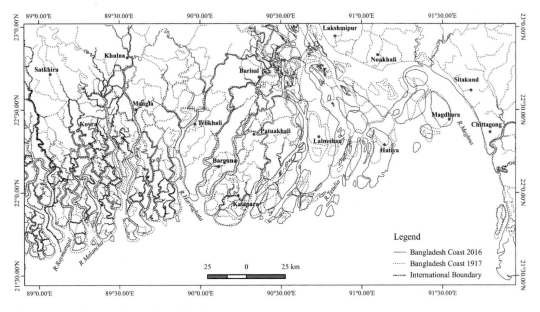

Fig. 8.8 Changing coastline of Bangladesh

growing by accretion. This is due to difference in sediment load carried by the Hugli estuary in India and the Meghna estuary in Bangladesh. Both the countries have extensively embanked rivers to protect the prematurely reclaimed land from ingress of saline water. The protection of coastal erosion may be a popular demand, but it is a task rather extremely difficult and too expensive. However, the planners must appreciate the fluvio-marine processes operating in the littoral tract of Bengal before adopting any management plan.

References

Bandyopadhyay S, Bandyopadhyay MK (1996). Retrogradation of the western Ganga-Brahmaputra Delta, India and Bangladesh, Possible reasons. In: Tiwari RC (ed) proceedings of 6th conference of Indian Institute of Geomorphologists, National Geographer, 31(1&2):105–128

Banerjee M (2008) Protnotatye Sundarban (in Bengali). In: Jana Debaprosad (ed) Sreekhanda Sundarban. Deep Prokashan, Kolkata

Bhattacharya A (2000) Embankments as large scale construction in the Indian Sundarbans and their impacts on the coastal ecosystems in Sundarbans. In: Guha Bakshi DN, Sanyal P, Naskar KR (ed) Sundarban Mangal. Naya Prokash, Kolkata

Brammer H (2014) Bangladesh's dynamic coastal regions and sea-level rise. Climate Risk Manage 1(2014):51–62

Brown S, And S, Nicholls RJ (2015) Subsidence and human influences in mega deltas: the case of the Ganges–Brahmaputra–Meghna. Sci Total Environ 527–528(2015):362–374

Chakraborty SC (2005) The Sundarbans-terrain, legends, gods and myths. Geograph Rev India 67(1):1–11

CSE (2009) Climate change/politics and facts. New Delhi

CSE (2012). Living with the changing climate/Indian. Sundarbans, New Delhi

Chattopadhyaya H (1999) The Mystery of Sundarbans. A. Mukherjee and Co. Pvt. limited, Kolkata

Goodbred SL Jr, Kuehl SA (1998) Floodplain processes in Bengal basin and the storage of the Ganges-Brahmaputra river sediment: an accretion study using ^{137}Cs and ^{210}Pb geochronology. Sed Geol 121:239–258

Halder G (2000) Western Sundarban: An Introduction. In: Guha Bakshi DN, Sanyal P, Naskar KR (ed) Sundarban Mangal. Naya Prokash, Kolkata

Hazra S (2010) Temporal change detection (2001-2008) of Sundarbans. Unpublished report, WWF, India

Hunter WW (1875) A statistical account of Bengal, 24 Parganas and Sundarbans, 1. Trubner and co, London

International Water Association (2017) Sundarban Joint Landscape Narrative (Unpublished Draft)

IUCN (2014) Bangladesh Sundarban Delta/Vision 2050. Bangladesh Country Office

Matin K (2016) Ujaner Bandh O Bangladesher Bhabishyat (in Bengali), Dibya Prakash, Dhaka

Oldham T (1870) President's address. In: Proceedings, asiatic society of Bengal for February 1870, Calcutta

Pargiter FE (1934) A revenue history of Sundarbans, Reprinted by W.B. Dist. Gazetteers (2002)

Paul AK (2002) Physiography of Mangrove Swamps-a study in the Sundarbans. In: Guha Bakshi DN, Sanyal P, Naskar KR, (ed) Sundarban Mangal. Naya Prokash, Kolkata

Ray NR (1949) Bangalir Itihas: Adi Parva, (in Bengali). Prachin Vanga Nirakasharta Durikaran Samiti, Kolkata

Rudra K (2011) The proposal of strengthening embankments in Sundarbans: myth and reality, www.counterviews.org/Sundarbans_aila.html

Rudra K (2012) Atlas of Changing River Courses in West Bengal. Sea Explorers' Institute, Kolkata

Rudra K (2016) Evolution of the Ganga-Brahmaputra Delta(Ganga-Brahmaputra bawdwiper bibortan (in Bengali). Sudhu Sundarban Charcha (4th year/4th issue), 8–16

Nady S, Bandyopadhyay S (2012) Trend of sea level change in the Hugli estuary India. Indian J Geo-Marine Sci 40(6):802–812

Sanyal P (1999) Global Warming in Sundarbans Delta and Bengal Coast. In: Guha Bakshi DN, Sanyal P, Naskar KR (ed) Sundarbans Mangal. Naya Prokash, Kolkata

Sarkar MH, Akter J, Rahman M (2013) Century–scale dynamics of the Bengal Delta and future development. In: proceedings of the 4th international conference on water and flood Management, pp 91–104

Unnikrishnan AS, Shankar D (2007) Are sea-level-rise trends along the coasts of the north Indian Ocean consistent with global estimates? Global Planet Change 57(3–4):301–307

WWF Report India (2010) Sundarbans: Future Imperfect. Climate Adaptation Report

WWF India (2011) Indian Sundarbans Delta: a vision. Policy Document

Flood in the GBM Delta

Abstract

Bengal is the ultimate outlet of 89,000 cumec water carried by the GBM system during the monsoon months. The flowing water often overtops the bank and submerges low-lying areas in the rainy season (June to September). The event is often described as disaster due to failure of protective measures to combat the hydrological extreme. The attempts of flood control, be it through building of embankments or by routing of flow from the reservoirs, did not work as expected. It is now understood that total freedom from flood is neither possible nor desirable. The strategy must switch over to preparedness-driven approach from ongoing relief-centric flood management.

9.1 Introduction

Bengal is historically flood prone due to its topography, location and hydrology. The Ganga and the Brahmaputra along with four major rivers (the Mahananda, the Teesta, the Jaldhaka and the Torsa) enter the Bengal basin through the Rajmahal–Meghalaya gap with huge volume of water and sediment load. The Barak which is headstream of the Meghna has a different entry route to Bengal through the Sylhet basin. The water carried by GBM system is 89,000 m^3/s during the monsoon, and it dwindles to 5714 m^3/s in the lean months (Matin 2016). The Meghna and its tributaries drain the north-eastern part of Bengal basin which receives highest rainfall in the world. A group of seven rivers, including the Damodar, drain the western upland of South Bengal and ultimately join the Bhagirathi–Hugli River. The peak monsoon discharge in the Ganga, the Brahmaputra, the Meghna and their tributaries often overtops the bank and submerges the low-lying tract. The deepening of channel and sediment dispersal are two major functions of the flood. The magnitude of flooding is governed by the many factors such as the duration and intensity of rainfall in the basin, volume and velocity of water, the slope and cross-sectional area of channel, the sediment load and nature of land-cover. The flood is often aggravated due to extension of railways and highways across the channels with an inadequate passage for floodwater. The most flood-prone areas of Bengal are plains of the North Bengal, Sylhet basin, plains adjacent to major rivers and coastal tract. A single-peak flood event is common in case of most of the rivers, but sometimes successive floods due to two or more consecutive cloudbursts may give rise to multiple peak events. Short-lived, but highly damaging flash floods are generally produced by intense rain during the July to September. The convergence of steep slopes with plains generates flash flood at both North Bengal Plains and Sylhet basin.

© Springer International Publishing AG, part of Springer Nature 2018
K. Rudra, *Rivers of the Ganga-Brahmaputra-Meghna Delta*,
Geography of the Physical Environment, https://doi.org/10.1007/978-3-319-76544-0_9

A riverine flood is generally caused by high-intensity precipitation within the catchment. The heavy rainfall in the Himalaya often causes landslides and the resultant debris-dam in the channel temporarily causes water to accumulate in excess of channel capacity. When the accumulated water either bursts or overtops the barrier, the huge volume of water flows down and may cause devastating flood. Such an event happened at Jalpaiguri in 1968 when a series of temporary dams across the bed of the Teesta were breached. The flood in the coastal areas is caused by a combination of tidal surges, cyclonic storms, hurricanes and tsunamis. Human-induced flood may occur due to the release of excess water from the dam, failure of hydraulic structures such as embankments or reservoirs and the impacts may be disastrous when the downstream regions are densely populated.

From a study of the flood characteristics of this subcontinent, eight major aspects have been identified and they are: (a) The Ganga–Brahmaputra basin is so aligned that its lower reach is hydro-geomorphologically vulnerable to large floods, (b) the monthly flow in all rivers of the GBM Delta are extremely skewed. The peak discharge in case of the rivers of South Bengal is generally observed in the late August or September but the flow in rivers goes to the peak about a month earlier in North Bengal. (c) The extreme flood events occur after longer interval and the second half of the twentieth century has experienced some most severe floods. (d) The flood magnitudes and frequencies in this subcontinent are primarily caused by the vagaries of the monsoon, (e) the natural and tectonic changes in river bed, shifting of river courses, breaching of man-made embankments are also the major causes of high-magnitude floods, (f) some people believe that there is very little scientific evidence to support the notion that man-induced environmental degradation in the Himalaya has directly aggravated the flood problems in the last few decades but (g) the proponents of the thesis of Himalayan Environmental Degradation claim that increasing deforestation and erosion in Himalaya have produced a huge sediment load which is transported downstream and deposited in the lower reach. This has caused an expansion of floodable area (Kasperson et al. 1999; Kasperson and Kasperson 2001). (h) The tidal fluctuation in the lower GBM delta does not allow easy downstream flow of flood water. (i) Total freedom from the flood through structural measures has proved futile. The flood control and management programmes have not been able to provide the expected relief in the most flood-prone areas.

The Ganga, the Brahmaputra and the Meghna systems generate extreme flow and huge suspended loads when moisture bearing south-west monsoon strikes the Himalaya (Subba 2001; Chapman 2005). The floods in Bengal like other third world regions *are interpreted as hazards not because of the intrinsic attributes of natural phenomena, but because of the failure of the socio-technical system to cope up with the situation* (Biswas 1988).

9.2 Changing Rainfall Pattern

The rainfall in this subcontinent is extremely skewed and more than two-thirds of annual rainfall is experienced in four monsoon months (Rudra 2015). The unusually concentrated rainfall due to cloudburst is the major cause of flood in Bengal. It is observed that the rain during the first part of the monsoon recharges the groundwater table and rejuvenates both rivers and wetlands. There is more runoff and less infiltration in the second half of the rainy season when a cloudburst may cause disaster. It is revealed from analysis of rainfall data of the period 1901–2010 in West Bengal that incessant rainfall causing the devastating flood may recur at an interval of more or less two decades. It is further revealed that the average monthly rainfall in the month of September in South Bengal and that in July in North Bengal have increased appreciably (Fig. 9.1). These are the major causes of increasing the magnitude of flood. The rainfall varies widely in Bangladesh. While the annual

9.2 Changing Rainfall Pattern

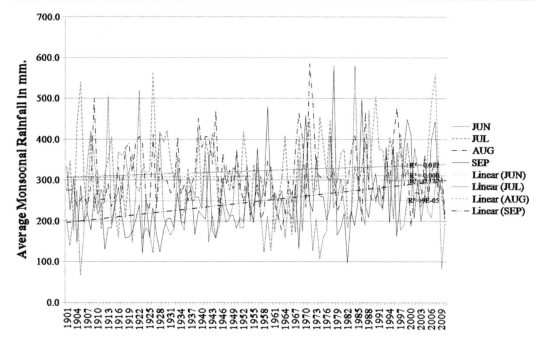

Fig. 9.1 Trend of monsoon rainfall in southern part of West Bengal (1901–2010)

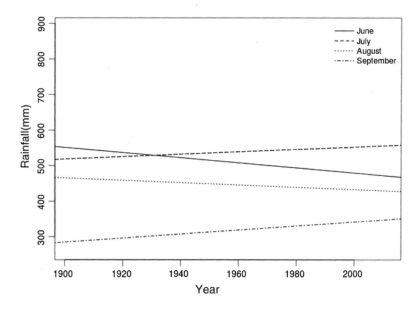

Fig. 9.2 Trend of the monsoon rainfall at Dhaka, Bangladesh. (1900–2010)

average rainfall is 2300 mm, it is only 1200 mm in the Barind Tract (area lying to the west of the Brahmaputra) and 5000 mm in the north-east part of the country. Historical data of rainfall recorded at Dhaka show a different trend. The monthly average rainfall has declined and increased alternately in four monsoon months (Fig. 9.2).

9.3 Location and Topographic Expression

The rivers draining Bengal have their sources either in the Himalaya and the hills along eastern border or in Chotanagpur Plateau. Both in cases of North and South Bengal River basins, the uplands and low lands are disproportionately distributed. The mountains or undulating terrain cover a larger area than the plains within a catchment. Whenever the rivers debouch on plains, they tend to be wide and sluggish. For example, the Damodar and its tributaries drain 20,874 km^2 and its lower 3700 km^2 is declared as flood prone. The southern part of the Meghalaya Plateau receives the highest rainfall in the world, making the Sylhet basin prone to flood. The same is true in the case of North Bengal Rivers like the Mahananda, the Teesta, the Jaldhaka, Torsa and Sankosh where mountains cover larger share of each basin compared to lower plains. The huge volume of water from a vast and intricate drainage network in the mountain is poured into a comparatively smaller plain segment of the basin where the main channel is wide and shallow. In such physiographic disposition, the flood is the inevitable hydrological event.

Every river of the delta has an active floodplain which is flooded frequently. The active floodplain of the Ganga is about 15–25 km wide at places while that of Bhagirathi is about 10 km (Figs. 9.3, 9.4, 9.5 and 9.6) The Jamuna has wider floodplain. The Meghna system gets a supply of water from the Meghalaya and mountains of Manipur and Tripura and causes flood in the downstream areas. There are some sub-basins locally called *bill* which accommodate rainwater and look like vast water bodies during the monsoon season.

9.4 Major Floods of Bengal

About 60% area of Bangladesh and 42% area of West Bengal are at risk from flooding. Entire Bangladesh except uplands of Barind, Maghupur and Chittagong Hills are subject to recurrent floods. The flood-prone areas in West Bengal have expanded continuously from the 17 m contour line in 1950s to 20 m in 1960s and to 26 m in the 1978 floods (Biswas 1988). The floodable area increased due to expansion of roads and railways which created drainage congestion. The official statistics of the floods that occurred during 41 years (from 1960 to 2000) show that only on five occasions the state has not faced any severe flood, when areas less than 500 km^2 were inundated. The marooned area crossed 20,000 km^2 in four different years and the flood of medium magnitude, i.e. 2000–10,000 km^2 occurred on 10 occasions. (Govt. WB 2011). The extremely high rainfall had always been the cause of major floods (Table 9.1).

In 1968, one of the most disastrous floods ravaged the entire lower Teesta basin and that was caused by exceptionally heavy rains in upper catchment. The debris of successive landslides in Sikkim created a series of temporary dams across the Teesta. The water accumulated beyond the debris and breached the barriers one after another and ultimately a huge volume of water rushed downstream to cause havoc. Nearly 1500 km^2 of cultivated area was submerged, 2000 human lives perished along with many livestocks and immovable property estimating Rs. 92 million were destroyed (Biswas and Nandi 1976).

In late September 1978, flood water submerged about 30,000 km^2 of South Bengal rendering 15 million people homeless. The estimated loss was Rs. 10,910 million, excluding the loss due to breakdown of communication and essential services. The hydraulic structures such as the dam on Hinglow, Tilpara barrages were breached and deposition of infertile sand made the left the bank of Ajay unproductive (Biswas 1988). In 2000, there were three major areas of flood. The first one was in Maldah. The second one was in the lower Damodar Valley. However, the biggest was the one that covered the four districts Birbhum, Murshidabad, Nadia and North 24 Parganas. That flood was described as the 'millennium flood' which caused distress to 20 million people in India and three million people in neighbouring parts of Bangladesh (Chapman and Rudra 2007).

In July 2017, six districts in the lower Damodar basin were marooned and North Bengal remained disconnected from the remaining part of the state for about a month. While more than 15 million people were affected, the total loss of crop and other assets amounted to INR 140,000 million.

9.4 Major Floods of Bengal

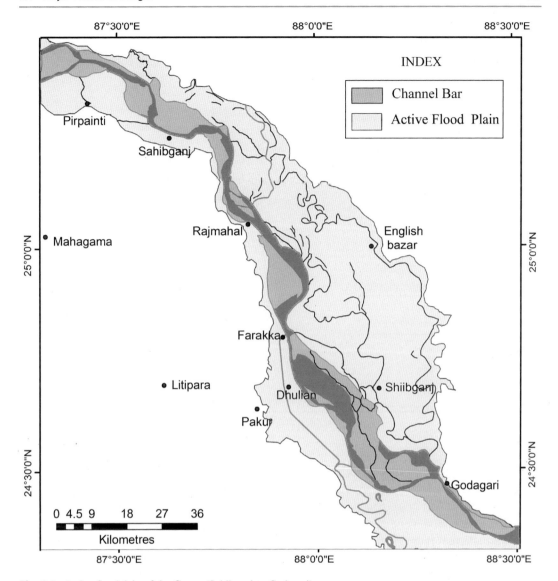

Fig. 9.3 Active floodplain of the Ganga (Sahibganj to Godagari)

The flood is an annual event in lower Damodar area. The areas which are flooded almost every year are Khanakul-I, Khanakul II, Amta, Pursura, Jangipara, Arambagh and Tarakeswar. When DVC-induced water synchronizes with rainwater of lower catchment and tidal backflow from the Hugli River creates a hydraulic dam intercepting free flow of water, the situation becomes grim. The role of DVC in flood management is discussed in Chap. 11. The southern part of West Bengal experienced flood in 1978, 1986, 1988, 1993, 2000, 2001, 2005, 2006, 2007, 2008, 2009, 2011, 2015, 2016 and 2017 (Chapman and Rudra 2007; GoWB 2009, 2011, 2017).

Three types of flood generally affect Bangladesh. These are (a) flood due to local intense rain; (b) that due to huge transboundary flow; and (c) cyclone induced submergence of coastal areas. Bangladesh (erstwhile East Pakistan) experienced disastrous floods in 1954, 1955, 1974, 1987, 1988, 1998, 2004 and 2007. The coastal tract of Bangladesh experienced severe

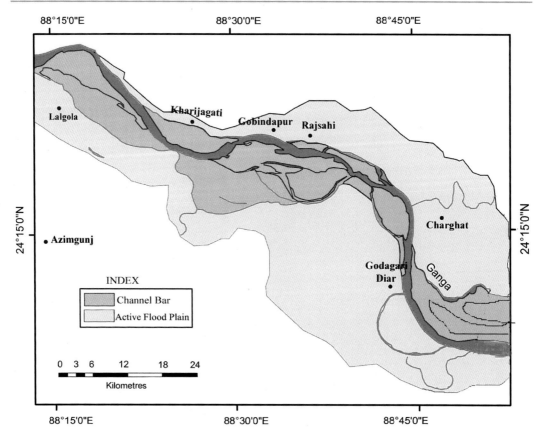

Fig. 9.4 Active floodplain of the Ganga (Lalgola to Charghat)

storm surge in 1942, 1950, 1970, 1974, 1988, 1991, 1994, 1995 and 2007. The country experienced most extensive flood in 1998 when 68% area was affected (http://www.ffwc.bd/images/annual14.pdf). In the last 100 years, the death toll due to the flood in Bangladesh is reportedly more than 50,000 and 32 million victims were homeless. The most disastrous event took place in 1970 when a tropical cyclone ravaged coastal Bangladesh causing the death of 500,000 people (https://en.wikipedia.org/wiki/1970_Bhola_cyclone).

9.4.1 Flood Management

The people of Bengal living on the floodplains learnt to live with flood from time immemorial. The farmers were aware of ecological benefits of flooding of their farmland as each flood used to replenish agricultural land with a layer of new silts. The farmers used to cut the bank of the river and allow silt-laden flood water to submerge the field. This was described as 'overflow irrigation' by Willcocks (1930). Every flood-prone area was agriculturally prosperous. Bardhaman District in lower Damodar basin could produce so much of crops that it ranked first in India (Hamilton 1820). Notably, Damodar basin is rich in coal reserve in its upper reaches and agriculturally productive in plains. Still the Damodar was described as 'Sorrow of Bengal', due to its recurrent flood which often delinked railway and highway. In mid-nineteenth century, the Damodar was embanked to protect GT road and East India Railway from the flood. Though it ensured protection against low-intensity flood, but also invited much ecological rapture. The Indian Mirror (quoted by Willcocks) reported on 21 December 1907—'*To protect the EI railroad from Burdwan onwards, a*

Fig. 9.5 Active floodplain of the Bhagirathi (Raghunathganj to Katwa)

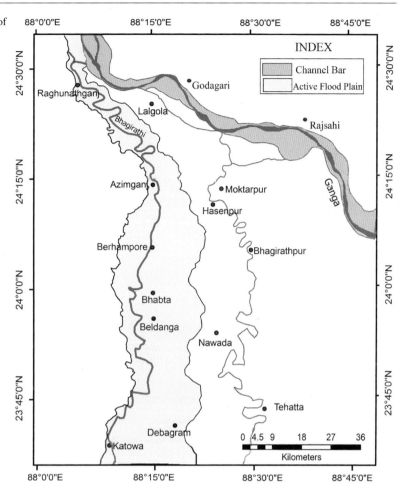

high embankment was constructed on the east bank of the Damodar to stop the floods. The PWD extended their operations to the other rivers …… *Wherever embankments were thrown up, tanks and artificial lakes were choked with weeds and crops suffered from want of fertilizing silts. Scarcity and famines have become more frequent and severe, and plague and fever are caused*'. The floodwater of the Damodar looks red and rich in nutrients. In fact, the embankment created horizontal disconnectivity between river and its floodplain. Even the distributaries which had been best possible outlets of the floodwater were beheaded (Fig. 7.6). The impact on agrarian economy was detrimental leading to the famine in 1865–66. Saha (1933) noted '*If there be anything like justice in the world, the people of Burdwan are entitled to compensation from parties concerned, for these terrible inflictions on them*'. The flood situation in Bengal was aggravated by the expansion of roads and railway which were built on high embankment intercepting flow of floodwater(Ray 1932). This increased the duration of the flood. It is important to note that the strong embankment built along left bank gradually pushed flood-prone area of the lower Damodar from the left bank to Ghatal and Khanakul area on right bank. In fact, considering priority to protect the GT road and railway, the right embankment was demolished in 1858 and allowed floodwater to rush towards the rural areas of Haora and Hugli Districts. The 'Sepoy Mutiny' (first war of independence) motivated colonial Government with a view to securing main communication lines from the flood so that army could reach the areas of revolt quickly. Further the uninterrupted transportation of coal from the mines of the upper

Fig. 9.6 Active floodplain of the Bhagirathi (Katwa to Kalyani)

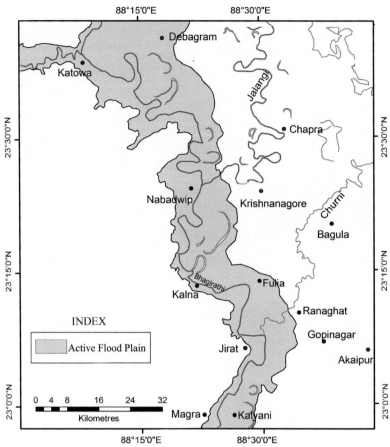

Table 9.1 Extreme monsoon rainfall (mm) in the southern part of West Bengal (1956–2000)

Days in September	1956 23–27 (Sept)	1959 8–12 (Sept)	1959 30–03 (Sep–Oct)	1978 31–03 (Aug–Sep)	1978 27–01 (Sep–Oct)	1995 26–28 (Sept)	2000 18–21 (Sept)	2017 20–26 (July)	Monsoon rainfall (mm)
Murshidabad	256	141	370			276	470	*	1167
Birbhum	321	121	206	102	636	347	800	221	1143
Nadia	296	203	331	87	349		311	*	1175
Bardhaman		229	231	76	162		369	249	1174
Hugli	276	291	274				352	233	1208
Haora	331		254				122	325	1241
Purulia		142	168	345	208		169	393	1163
Bankura		201	202	281	417		50	470	1160
Medinipur (E and W)	133	193	192	351	248		90	281	1202
24 Parganas (N and S)	284	276	298	154	646	151	160	*	1249

Source Chapman and Rudra (2007). Irrigation and Waterways Department, Govt. of West Bengal, and unpublished records of the Indian Meteorological Department
*No data available

Damodar basin down to port of Kolkata was another important issue under consideration. The best technical solution for flood management could have been proving wider space to the river between the embankment and routing water through old the distributaries.

The British left India in 1947 and Bangladesh was separated from Pakistan in 1971. The task of flood management was given to some engineers who continued to rely on same style of hydraulic engineering. To date, more than 10,000-km-long embankment in West Bengal is maintained by Irrigation and Waterways Department. In Bangladesh, the total length of the embankment is reportedly 7555 km. In the post-independence era, some dams/reservoirs were built with twin purposes of flood control and irrigation. The reservoirs have limited capacity of storing water and rapid sedimentation led to reduction of their capacity further. It has been observed that reservoirs were compelled to release water in the years of excessive rain and that caused dam-induced flood in the lower reach (Govt. of WB 1959).

It is quite clear that now is the time to reopen the debate in Bengal about what constitutes appropriate and sustainable development. There are alternatives to the paradigm of engineering defence and control. Perhaps urban areas do need some flood defence, but for the vast majority of the country, a reversal to 'normal' flooding could be beneficial. Existing road and protection embankments could be breached at chosen sites at fairly frequent intervals. As flood rises, the water enters new territories in an observable fashion from selected points. The problems of lack of warning, rapidity and unpredictability are eliminated. As a further precaution, there has to be a programme to aid the introduction of at least one substantial flood shelter in each village or hamlet. This could have private uses in normal times, providing rent to offset its cost. In addition, the de-silting and maintenance of selected waterways can change rural communications. In short and in general, these are proposals which open up the question of what constitutes appropriate and sustainable development in the Bengal Delta.

9.5 Issues of Disaster Management

Administrative set-up for dealing with disasters in India dates back to 1871, when the British set-up the Department of Revenue and Agriculture and Commerce to deal with the frequent famines. Famine Commissions were set-up and codes for famine relief formulated. In 1881, the Department of Agriculture was constituted on recommendations of the Famine Commission, and upgraded to Ministry of Agriculture in 1947. By the mid-1970s, this Ministry became the nodal centre for dealing with floods and droughts which had maximum and recurrent impacts on the agrarian economy. Earthquakes and cyclones, being random, were to be dealt with as and when they struck (Kapur et al. 2005).

At the state level, while no major measures have been taken in the past for cyclones, landslides and earthquakes, attention of the Government right from the days of the British Raj has been on floods and droughts, due to implications for agriculture and thereby, food supply. West Bengal Government professes that 42% of the area and nearly 148 of the 341 community development blocks in the State is flood-prone. Jalpaiguri and Koch Bihar, being located at the foot of the Himalayas, present the 'flash flood' scenario. The entire Bangladesh except the highlands of Barind, Madhupur and Chittagang is under the risk of flooding.

The response to floods has been twofold:

- Structural measures such as embankments in the colonial phase, followed later by dams for flood control.
- Relief in terms of food, cash compensation and sometimes seed distribution after the disaster.

Thus, one finds 10,539 km stretches of river embankments, and eight dam-barrage systems ostensibly for flood control, irrigation and electricity generation for West Bengal. About 24% land area and 39% of net cultivated area of Bangladesh is declared as protected by 7555-km-long embankment.

Understanding that total freedom from flood is not possible; the present approach could be divided into three core components

- Strengthening early warning of flood,
- Strengthening preparedness of family, community, institutions and the Government to face disaster,
- Strengthening disaster mitigation aspects in the long-term development.

9.5.1 Early Warning

For certain hydro-meteorological hazards, such as cyclones, droughts and floods, systems for issuing early warnings exist. The Central Water Commission in association with Irrigation Department, the Government of West Bengal is authorized to anticipate floods; the Indian Meteorological Department is authorized to do likewise for meteorological droughts and cyclones. Rainfall gauges, river gauges, wireless sets, dam discharge data, cyclone tracking satellites are in place for these purposes. Models exist to interpret the data and predict the next course of action. The Bangladesh Water Development Board is mandated for forecasting through 54 stations widely distributed in Bangladesh.

A gap lies in communicating information and data to the vulnerable community at the grassroots level so that it is of use to them. This calls for a rethinking on information dissemination and use mechanisms. The present framework considers that information is to be used only by the administration whereas experience all over the world have shown that information dissemination and use to and by the local community actually reduces vulnerability and greatly improves preparedness. Recent action research on early warning has come out with detailed recommendations that may be referred to in this context.

9.5.2 Strengthening Preparedness

Strengthening preparedness of family, community, *panchayat*s and administration is perhaps the most significant approach in recent times. Born of international experience, it refers to measures for improved understanding and collective confidence and readiness of the family, community and that of administration to face a disaster. Preparedness has three main aspects:

- Family-level preparedness by which, the family takes critical measures such as stocking food and other commodities, constructs elevated shelter (a raised wooden platform), keeps a stove or clay oven, elevates platforms of tube wells for drinking water, applies bleaching powder and alum to ensure safe drinking water and other such measures depending on situation.
- Community preparedness includes to creation of a group of trained volunteers to facilitate warning dissemination, rescue of victims, water-sanitation, shelter management, waste disposal. And pre-storage of medicine and dry food, and boats to transfer the victims.
- Institutional preparedness refers to specialized rescue forces (professional rescue teams, rescue boats, expert drivers, rescue equipments), pre-positioning of large stocks of food, medicine, anti-venom medications for snake and insect bites, emergency medical team, emergency transport vehicle, diarrhoea control measures, emergency engineering team for road and bridge repair, ambulances, boats, emergency communication equipment, route maps, preparedness of staff.

9.5.3 India's National Policy on Floods

The first National Flood Management Policy was published in January 2008 with a view to '*minimize vulnerability to floods and consequent loss of lives, livelihood systems, property and damage to infrastructure and public utilities*'.

The National Remote Sensing Agency and Indian Space Research Organisation have been entrusted with the responsibility of identifying and mapping flood-prone areas of the country. It has been realized that structural measures are not enough to prevent flood, nor is it possible to achieve complete 'freedom' from flood. This has paved way switch over from a relief-centric to a preparedness-driven approach.

All state Governments have been advised to prepare floodplain zoning map and scientific land use plans of vulnerable areas. The Government also endeavours to develop international collaboration with neighbouring countries towards exchange of hydro-meteorological data. Flood management in West Bengal has been dependent on the colonial legacy of embanking rivers and the post-independence reductionist engineering approach of building dams. In both cases, the plan to combat hydrological extremes of flood and drought was made in isolation from sediment management programmes. The sediment trapped between embankments and upstream of dams/barrages ultimately caused the decay of the drainage system and expansion of floodable areas. The lesson from the past is that the volume of water generated during a late monsoon cloudburst can neither be totally accommodated in reservoirs nor contained within embankments. A regulated spill-over of sediment-laden flood waters can become a sustainable option to restore floodplain ecology. Moribund channels can be resuscitated by selective dredging to make best possible outlets of flood water.

9.5.4 Policy in Bangladesh

Two consecutive floods in 1954 and 1955 prompted erstwhile the Pakistan Government to constitute a committee to explore the remedial measures and to formulate a policy for flood management. The initial thrust was on building the embankments and the creation of polders (Aktar 2003). But adverse impacts of such structural measures were realized. Subsequently, several non-structural measures of flood management were adopted. Such programmes include building new structures above highest flood contour; schools were planned to build in such a way that those could also be used as flood shelters. Besides, encroachment on the floodplain was regulated as far as possible.

It is now understood that keeping the huge monsoon flow within bank to bank of the river is impossible. The only option is managing to live with flood and this can only be achieved integrating the structural and the non-structural measures of flood management.

9.5.5 Linking Flood Management with Development

Perhaps this is the most crucial aspect of disaster management today—an aspect that has several ramifications. At one plane, it seeks to redefine our standards for construction measures such as roads, railways, buildings (especially the multi-storied ones), electric installations, communication networks, schools, hospitals and other critical infrastructure, bearing in mind aspects of disaster risk. The second arena is to work to improve the overall environment situation. The third aspect demands looking at livelihood options of the people. Sporadic works need to be integrated into a State Disaster Management Policy.

West Bengal and Bangladesh need collaborative partnership between public authorities and the civil society to minimize risks of disaster. The focus needs to be on the community which faces the real brunt of disasters. It is, indeed, possible to minimize the risks and impact of disaster if we act together in all earnestness. The way forward is to build a collaborative flood management policy, where both the Governments will commit themselves to supplement and compliment each other's roles and responsibilities.

References

Aktar ANH (2003) Bangladesh: flood management downloaded on 28.10.2016 (https://cleancookstoves.org/binary-data/RESOURCE/file/000/000/78-1.pdf)

Biswas A (1988) To live with flood : the case of West Bengal. In: Paper presented in the international symposium on river bank erosion, flood and population displacement, Jahangir Nagar University, Dhaka

Biswas A, Nandi H (1976) Floods in Eastern India. Eastern Regional Affairs (ERA) 1:23–30. Calcutta

Chapman GP (2005) Natural and human environment of the Ganges-Brahmaputra-Meghna Basin. In: Subedi SP (ed) International watercourses law for the 21st century. Ashgate, UK

Chapman GP, Rudra K (2007) Water as foe, water as friend: lesson from Bengal's Millenium Flood. J S Asian Dev 2(1):19–49

Govt. of Bangladesh (2014) Annual flood report, 2014. http://www.ffwc.bd/images/annual14.pdf

Govt. of West Bengal (1959) Final report of the West Bengal Flood Enquiry Committee, I

Govt. of West Bengal (2009) Unpublished records on flood damage. Irrigation Deptt

Govt. of West Bengal (2011) Unpublished base paper related to the activities of the newly constituted Flood Control Commission

Govt. of West Bengal (2017) Unpublished Records on Flood Damage

Hamilton W (1820) Geographical, statistical and historical description of Hindostan, vol I. London

Kapur A et al (2005) Disaster in India/studies of grim reality. Rawat Publication, New Delhi

Kasperson RE, Kasperson JX, Turner BL (1999) Risk and calamity: tracjectories of regional environmental degradation. Ambio 28(6):562–569

Kasperson JX, Kasperson RE (2001) Global environmental risk. United Nations University Press, Tokyo

Matin K (2016) Ujane Band O Bangldesher Bhabisyat (in Bengali). Dibya Prakash. Dhaka

Ray PC (1932) Life and experience of a Bengali Chemist, II, pp 159–160. Reprinted by Asiatic Society (1996), Kolkata

Rudra Kalyan (2015) Paschimbanger Jalsampad/Sankater Utsyasandhane (in Bengali). Sahitya Samsad, Kolkata

Saha MN (1933) Need for a hydraulic research laboratory. Reprinted (1987) in Collected works of Meghnad Saha. Orient Longman, Calcutta

Subba B (2001) Himalayan water: promise and potential: problem and politics. Panos, South Asia, Kathmandu

Willcocks W (1930) Ancient System of Irrigation in Bengal, University of Calcutta

Management of Rivers in the GBM Delta

Abstract

The vast alluvial plain, bright sunlight and plenty of silt-laden water have been the cardinal factors making Bengal agriculturally prosperous and ecologically productive. The farmers of Bengal had been living with the flood through ages. But the colonial rulers and landlords having no understanding about the ecological services of the flood started to embank the rivers and that invited serious drainage congestion. The situation was further aggravated when roads and railways were built with inadequate outlets of flood water. The second half of the twentieth century witnessed building of dams/barrages across many rivers with the targets of flood control, irrigation and hydro-power generation. Though these projects have some positive impacts, the ecological raptures in the basins were alarming. The rivers were horizontally disconnected from their floodplain due to building of embankments along the banks and also longitudinally disconnected by dams and barrages.

10.1 Colonial Period

The vast natural resources of Bengal lured the Europeans to invade this part of the country regularly since the sixteenth century. Bengal had the long tradition of producing varieties of rice, fine cotton, silk and indigo. The agrarian prosperity was even much better than that of Egypt. The sediment carried by the Ganga and its tributaries had been rich in nutrients and replenished the fertility of the soil during the flood. Tavernier, the French traveller who visited Bengal in 1666 wrote:

> Egypt has been represented in every age as the finest and most fruitful country, and even our modern writers deny that there is any other land so particularly favoured by the nature: but the knowledge I have acquired in Bengale, during two visits paid to kingdom, inclines me to believe the pre-eminence ascribed to Egypt is rather due to Bengale. The latter country produces rice in such abundance that it supplies not only the neighbouring but remote states (Crooke 1925).

Egypt is said to be the gift of Nile, and Bengal is that of the Ganga. Bengal, which has been drained by an intricate network of rivers, is inherently prone to recurrent floods. The flood here is not an evil but a blessing in disguise. The diminishing fertility of soil is replenished by the flood. Ayeen Akbery or The Institute of Emperor of Akbar noted that:

> Most rivers of Bengal have their banks cultivated with rice, of which there a variety of species. The soil is so fertile in some places, that a single grain of rice will yield a measure of two or three seer. Some lands will produce three crops in a year. Vegetation here is so extremely quick, that as fast as the water rises the plants of rice grow above it, so that ear is never immersed. (Gladwin 1800).

Being located at the tail end the Ganga–Brahmaputra basin covering 1.6×10^6 km^2, and the Bengal plain renders passage to the sea 631 bcm of water annually. This water carries suspended sediment load amounting 1667×10^6 ton (Milliman and Meade 1983). The water flowing through the rivers of Bengal fluctuates due to temporally skewed rainfall. Moreover huge transboundary water flows through the rivers of Bengal towards the sea. More than 80% of the yearly flow passes during June to September. The bank-full water carrying huge sediment load flows through the rivers during late monsoon. The excess water often overtops the banks and causes a flood. Indeed, farmers of Bengal always reaped the benefits of flooding which renewed fertility of soil through deposition of nutrient-rich sediment on farming land (Willcocks 1930). The silt-laden waters were taken to the agricultural field through artificial cuts in banks of rivers, and these were known as *kanwa* meaning 'to dig' in old Persian or Arabic language. The muddy water that spilled over the floodplain used to renew the fertility of soils and also prevented spreading of malaria. Willcocks opined that the moribund rivers of Bengal, especially in lower Damodar plains which are now called *kananadi* were old irrigation canals.

Willcocks argued:

1. *The spacing apart of the canals is just where canals would be placed to-day if there were none already on ground*
2. *They are fairly parallel and continuous in the direction in which they start, which is absolutely artificial*
3. *They are wide and shallow to carry the beneficial muddy surface waters of the rivers and avoid harmful sandy waters of the beds*
4. *The villages are constructed on their banks as villages would naturally be constructed, under the conditions in Bengal, on raised bank*
5. *All the canals were originally dug straight as a matter of course, but their winding courses to-day are true gauge of the friability of the soil they traverse. Their winding courses along their original alignments are nature's masterful handiwork.*

One can put forward many arguments against the hypothesis stated above but cannot deny that they, either as canals or natural distributaries, used to distribute sediment-laden water over the farmland. That was congenial to traditional rice cultivation in Bengal (Biswas 1981). Such traditions declined during the colonial, when the traditional irrigation system was replaced by modern irrigation engineering (ibid.). The British takeover of Bengal happened at a time when invention of steam power initiated the revolution in navigation and railway (Chapman and Rudra 2007). This was the beginning of the era of command and control over nature. This paradigm shift in the technology resulted in the long run large-scale ecological degeneration causing unbearable sufferings of the farmers. This was described by Bakim Chandra Chattopadhyaya in his write-up *Bangadesher Krishak* (Farmers of Bengal) *in 1873*.

10.2 Altered Hydrological Regime

The colonial hydrology of India has been discussed by many authors. While explaining the relationship between colonialism and water, De'Souza (2006) identified three overlapping but discrete areas of concerns. These are colonial irrigation strategies, decline, elimination and appropriation of traditional irrigation technologies and hydraulic endowment. The last area of concern encompassed the colonial attitude of river management in respect of flood, inland navigation and holistic river management. The British Engineers had apparently the pious intention to develop a network of canals so that farmlands can be irrigated even during lean months. They needed to store monsoon water and transfer the same to non-monsoon season. A long canal connecting Kansai in Medinipur to Damodar at Uluberia was excavated with the twin purposes of irrigation as well as navigation. It was opened for traffic in 1863. A weir on

10.2 Altered Hydrological Regime

Kansai at Medinipur was built in 1869 to induce water in the canal and another at Panskura was in process of making. But the latter washed away during the flood in 1872. The canal was designed without any prior estimate of the seasonal variation of flow in river and the area that could be irrigated. So there was a wide breach between the target of irrigated area and that actually irrigated. The canal with high banks created drainage congestion and the dredging of main drainage lines and building of sluices on them was proposed (Inglis 1909). While the Kansai-Damodar link canal was excavated to facilitate irrigation, the two other along the coasts were designed for navigation.

Ray (1932) noted-

> … In the pre-railway days, in order to facilitate communication between Calcutta and Puri (Temple of Jagannath), a canal was cut (Hijli Tidal Canal-1868-73 and Orissa Coastal Canal 1880-86). This canal runs parallel to the sea-coast in many portions of Tamluk and Contai and is protected by high embankment on both sides. During the excavations of the canal, many outlets for passage of water, which ran athwart the natural drainage, were filled up.

The Permanent Settlement Act (1793) introduced the feudal system that was known as the *Zamindari* system in Bengal. The right of land was given to landlords who agreed to collect revenue from farmers and pay a portion to the British rulers. Since then, many rivers in Bengal were jacketed with a vision to combat recurrent flooding. This was done without any regard to the fluvial dynamics of the delta. The rivers which were initially attempted to bring under control included the Kansai, the Silai, the Dwarkeswar, the Ajay, the Mayurakshi and the Damodar. The embankment initially ensured relief from low-intensity flooding but led to the decay of the fluvial system aggravating waterlogging. Mahalanobis (1927) opined *'Embankment in the low-lying tract might for time being prevent overflow from rivers but would tend to raise the bed of rivers still further, and thus made the situation much worse in the long run'*. Majumdar (1941) rightly noted that embankments intercept the sediment dispersal on the floodplain and ultimately cause the decay of the river. The temporary benefits derived at the first phase gradually disappear, and the cost of ecological degeneration was externalized on the next generations.

The embankment derived some initial benefits but the ultimate detrimental impacts were huge. The sediment load which earlier used to spillover floodplain was trapped between the embankments and resulting rise of the bed level and reduction of the water retention capacity. The embankment did not prevent high-intensity flood due to frequent breach and collapse. The sense of protection against flood ensured by the embankment ultimately proved to be myth. Neither the *Zamindars* nor British Engineers could foresee the hydro-geomorphological impacts of embanking the rivers in the long run. The programme of taming the rivers gradually extended from the plains of North Bengal down to the Sundarban. The rivers of North Bengal debouch on plain and form triangular depositional feature known as fan. These rivers are shallow and wide in this stretch. These rivers were also jacketed but flood continued to imperil the human society. The ecological degeneration relating to embankment building was manifolds but most important was that the farmlands were deprived from the annual deposition fertilizing silt. This was described as 'the red-water famine' in *Rarh* Bengal (Mukherjee 1938). The Bardhaman district was so gifted with the silt of the Damodar that it ranked first in agricultural productivity in the whole of Hindustan (Hamilton, 1815 quoted in Mukherjee 1938).

Impact of Railways

The condition deteriorated further as the standard European railway-road model of development was implanted on the inappropriate physical condition of Bengal and continued to expand further (Chapman and Rudra 2007). The control over Bengal by the British after the battle of Plassey in 1757 was contemporaneous to the Industrial Revolution in Britain. The availability of cheap coal, iron/steel and invention of steam power set the stage for railway revolution since 1830s. The investment in railways was huge but thought to be secured. There was a long debate in Britain over the comparative advantage of railway or canal. The engineers who advocated for improvement of navigation in Britain were blamed as suffering from *canal mania*.

Considering the intricate network of rivers, a group of engineers, especially Sir Arthur Cotton, advocated for development of navigation in Bengal which was then one of the main sources of raw materials for industrial growth in Britain. The rail and road lobby was ultimately successful to convince the policy-makers because of their stronghold in the British Parliament. The commencement of the coal mining since 1820 at Raniganj (Bardhaman district) area was one major cause of subsequent expansion of railways towards that direction. The Sepoy Mutiny or the first war of independence in 1857 also generated an idea that roads and railways would be best means of communication in India to deploy army quickly to the places of revolt. The dense tropical forest of the Chotanagpur Plateau was indiscriminately cleared to supply sleepers under railways. This caused fast erosion of topsoil from the undulating terrain and deposition of increasing sediment load in rivers. The roads and railways were built on high embankment to ensure uninterrupted movement even during the flood. The earth materials for building embankments were borrowed from ditches parallel to the lines of communication, and those borrow pits subsequently became pool of stagnant water and breeding grounds of malarial mosquito. The spread of malaria followed the paths of railway and highway. The construction of railway from Sealdah to Lalgola created a drainage congestion in the area between eastern bank of the Bhagirathi and the railway embankment causing widespread epidemic of malaria in Murshidabad and Nadia (Samanta 2002). Since the railways and highways in both north and south Bengal were aligned over the rivers and were built with inadequate bridges/culverts, there were drainage congestions, expansion of flood contours and decline in agricultural productivity. So the embankments, roads and railways were described as *satanic chains* (Willcocks 1930).

In fact the cost of urban-industrial living was externalized on rural Bengal. Prof. Megnad Saha (1933) in an article entitled *Need for a hydraulic Research Laboratory* explained the ecological degeneration of the lower Damodar plain since East India Railway connected upper India with Kolkata. Two years after opening of the railway in 1859, malarial epidemic caused deaths of about one million people in Hugli district and the fertility of soil declined in both the Bardhaman and Hugli.

The case of North Bengal was not an exception. Prof. Saha further noted that:

> The railway authorities in North Bengal build railways with insufficient waterways. These are responsible for devastating floods and the outbreak of malaria in North Bengal as well as for the fall in fertility of soil.

Prof. P. C. Ray (1932) criticized the railway authority for aggravating the flood. After the devastating flood of 1922 in North Bengal, he wrote:

> The Government is criminally and wilfully responsible for the great havoc….unless the narrow culverts were replaced by bridges of long span they would always be liable to the calamity of flood. And this is what exactly happened. The fact is that railway lines are always constructed with an eye to the interest of foreign shareholder. The less the cost, greater the expectation of dividend.

Mukherjee (1938) was also critical against about the road–railway in low-lying tract of deltaic Bengal. He opined:

> A systematic policy of road and railway construction in the eastern districts of Bengal would be a repetition of the mistakes, which have contributed in no small measure to the economic decline of the central and western Bengal. More attention should be diverted to the policy of the improvement of waterways and inland navigation, the making of new waterways by means of cuts, where none exists at present, the easing of bad bends of rivers and clearance of aquatic weeds in waterways. The German policy of 'railway and waterway' rather than 'railway versus waterway' is nowhere so indispensable as in the Eastern Bengal, where water-borne traffic is still one of the largest in the world.

10.2 Altered Hydrological Regime

The Damodar had a series of distributaries, namely Banka, Behula, Gangur, Kunti, *Kana* Damodar and *Moja* Damodar. Presently, the Damodar discharges its water through Mundeswari and Amta channel. But palaeo-channels are still most convenient outlets of floodwater. The railway connecting Haora with Bardhaman intercepted the passage of flood waters of the Damodar towards Hugli River. In 1853, a high embankment along the left bank of the Damodar from Jujuti down to its outfall was built. The primary objective of this construction was to protect the G. T. Road and railway connecting North India with Kolkata. Thus, the large part of the Damodar delta was trapped between two embankments built with the two different purposes, one for aligning railway above highest flood level and the other for protecting the same from high flood. The latter caused disconnection of all lower Damodar distributaries from their feeder. Another embankment built earlier along the right bank was removed in 1853 to render a passage to the flood water towards rural areas of Bardhaman and Hugli district. In 1855–56, the left embankment was breached near the off-take of Kana or Kunti River and the gap was left open till 1862. In 1863, the breach was plugged and the road leading to Memari was built over it and this again sealed the fate of both Kunti and *Kana Nadi*. The peasantry faced serious crisis leading to the outbreak of a malarial epidemic in 1864 and terrible famine 1865–66.

10.2.1 Structural Interventions

A remedial measure was planned in 1874 when the Jamalpur regulator was built to induce water in *Kana* Nadi (see Fig. 6.5). This was followed by construction of a sluice at Jujuti in 1881 to induce water in Banka River which was again intercepted at Kanchan Nagar by the construction of an anicut and from here the 30 km-long-Eden was excavated in 1881. This canal connected Kanchan Nagar anicut with Jamalpur regulator. Thus, a complex structural system was built to induce water in a distributary which was earlier naturally replenished. Thus, the colonial rulers altered the hydraulic regime of the lower Damodar area and introduced a structurally controlled irrigation system. The age-old irrigation system or overflow irrigation ensured the spreading off silt-laden and nutrient-rich water for agricultural land. The newly introduced canal system used to distribute relatively silt-free water withdrawn from superficial layer of flow. Before any such intervention agricultural productivity in lower Damodar plain was so high that Bardhaman district ranked first in agricultural production within India. But embankment caused such ecological degeneration that from the second half of nineteenth century, Bardhaman was neither productive nor healthy. From 1862 to 1872 about one-third of the population died from malaria which was locally described as the *Bardhaman fever*.

It must be emphasized that the river was the major backbone to the agricultural system, and the people had adapted themselves to the floods. Before the advent of the British, the main course of the Damodar shifted south-west and the east-flowing distributaries like the Banka, the Behula and the Gangur and south-east flowing the *Kana* Damodar and the *Moja* Damodar were beheaded. But these moribund channels played important role in the agrarian economy by dispensing silt and recharging groundwater pools during the flood. The prolific breeding fishes in the wetlands had been a natural control of malarial mosquito. But the flood control embankment intercepted these channels and their ecological services were ceased forever.

The life support channels gradually became stagnant pools of water and a great threat to public health. In 1872, Col. J. F. Stoddart, the superintending engineer of the south-western circle noted that *the silting up of the drainage channels, more especially the important ones, namely Kana or Kunti Nuddee, the Kana Damooda and the Surosuttee, is the principle cause of malarious fever*. Without any regard to the role of these distributaries in flood management and irrigation, *Kana* or Kunti River was dammed across its head in 1853 (Inglis 1909). This along with the left embankment of Damodar from Jujuti to its outfall sealed the fate of all palaeo-channels draining the trans-Damodar area. Thus, the entire area was trapped between the flood control embankment in the west and the railway embankment in the east. This great threat to public health and collapse of natural irrigation system caused a repercussion on the people and the British Government on the advice of Mr. Whitfield decided to construct a sluice at Jujuti to induce 200 cusec of water into Banka channel and then through a 33-km-long-Eden canal from Kanchan Nagar to Jamalpur where it flowed into the Kana and Kunti channel. Prior to excavation of the canal in 1787, cultivators used the waters from the *Kana Nadi* free of cost but as notified under the new Irrigation Act, the farmers were allowed to take water on payment. The hydrological flow was further altered by the construction of the Anderson Weir at Rhondia in 1933 with the objective of the expansion of irrigation.

The British Engineers except a few like Willcocks (1930) or Inglish (1909) had an incomplete understanding about the hydro-geomorphology of the Bengal Delta where farmers evolved practices to make best possible utilization of the natural system. The British hydraulic system failed to comprehend the significant role of the silt-laden flood water in Bengal agriculture. The structural intervention across the rivers, canals excavated to withdraw water and the building of linear flood control embankment created discrete controlled systems which impaired the delicately balanced and integrated river system of Bengal. The structures built across the Damodar at Rhondia or Kansai at Medinipur were low and rendered the flowing water the opportunity of overtopping the barriers. But the post-independence era witnessed massive structures which often left the downstream stretch of river totally dry. It was understood late that many ecological services were ceased subsequent to the taming of the rivers.

10.2.2 Pre-Mature Land Reclamation

The merchants from Britain assumed the administrative power of Bengal after the battle of Plassey under the leadership of Lord Clive in 1757. James Rennell was directed to conduct the survey and prepare maps to have a geographical knowledge of the territory in 1764. The land building in vast littoral tract of Bengal was then incomplete and had been submerged regularly during high tide. The area was drained by complex network of tidal creeks and covered with dense mangrove. While land remained 1.50–3 m above mean sea level, the tide might achieve heights between 5 and 6 metres. The occasional storm surge might be higher. In 1770, it was decided to clear forest from 54 islands in phases and polder the land to facilitate agriculture and settlement. The islands lying further east were untouched and remained covered with dense forest. The reclaimed areas were divided into several plots, and right of cultivation was given to some landlord who agreed to pay annual revenue to the East India Company's exchequer. The embankments disconnected the floodplains from the creeks, and sediment dispersal during high tide was impaired. Notably, tidal fluctuations twice in 24 h may cause accretion of five centimetre thick sediment layer. But this natural process was interrupted due to embanking of creeks. Alongside, the delta building continued in the non-reclaimed part of Sundarban where spillover of silt-laden water was not interrupted. That part now stands at a relatively higher level than its western counterpart. The destructive wave often breaches the embankment and also overtops at many places. The water remains stagnant for several days. The problems of Sundarban are discussed at length in Chap. 8.

10.3 Post-Colonial Period

The power and water are two important components of the modern civilization. While water has no substitute, the hydro-power is generally treated as green energy though it is strongly debated. Many large and run on river dams were built across Indian rivers to facilitate either irrigation or power generation. The interception of the rivers and construction of reservoirs were regarded as the most convenient method of water storage. More than five thousand dams built so far have submerged extensive forests and homelands of at least 40-million people who are economically backward. The benefits of projects, be it irrigation or hydro-power, were enjoyed by the people living in the lower catchment.

Former President of India K. R. Narayan while delivering the speech on the Republic Day, (2001) said

> 'Let it not be said by future generations that the Indian Republic was built on the destruction of the green earth and innocent tribals who have been living there for centuries'.

Jawharlal Nehru, the first Prime Minister of India described large dams as the *'temples of modern India'*. But he spoke otherwise in a meeting held November 17, 1958 and lamented that ' we suffered from the *disease of gigantism'*. Finally he favoured smaller irrigation projects. But modern India continued to rely on large multipurpose projects. The fierce debate over the efficacy of large dams its ecological sustainability divided the scholars into two groups.

India achieved long-awaited independence in 1947, and Bengal was divided into two unequal parts. The western part was included in India as a State which was named West Bengal while its eastern counterpart was named as East Pakistan. The latter subsequently emerged as Bangladesh in 1971. The Indo-Pakistan border was so delineated that the 54 rivers including the Ganga became transboundary and that set the stage for conflict over the sharing of flowing water. The partition of Bengal led to the influx of population from erstwhile East Pakistan to India and the demand for food increased manifolds. The obvious compulsion of the Government was to increase the food production which needed expansion of the irrigated area. The first agricultural revolution in early 1960s was executed with the introduction of high yielding seeds, chemical fertilizers, pesticide and large dam technology. The first multi-purpose river valley project was initiated in West Bengal under the aegis of Damodar Valley Corporation.

10.3.1 Damodar Valley Corporation

The Damodar is the most important tributary to the Bhagirathi-Hugli River. It originates from Palamou hills (600 m above MSL) of Chotanagpur Plateau and travels south-eastward for about 540 km and ultimately discharges into the Hugli River through Mundeswari and Amta channel. The former finds its outlet through Rupnarayan estuary and the latter discharges directly into the Hugli River near Shyampur of Haora district. In 1970, a shorter outlet was excavated to connect the Damodar with the Hugli and a lock was constructed across it at Garhchumbak to intercept the reversal of flow during high tide (see Fig. 7.6).

The Damodar drains Hazaribag and Dhanbad districts of Jharkhand before it reaches Dissergarh. The lengths of Damodar and Barakar above the confluence are 312 and 238 km. respectively while the catchment areas of these two tributaries are 10,887 km^2 and 6869 km^2, respectively. The area lying above Dissergarh may be called upper catchment. From here, the river flows south-eastward for length of about 145 km and takes a southward turn at Jamalpur of Bardhaman district. This part (Dissergarh to Bardhaman) is identified as the middle reach. While the 3700 km^2 area of the Damodar delta below Jamalpur may be called the lower reach (see Fig. 7.5).

Excluding the lower Damodar distributaries and taking into account two major outlets—the Mundeswari and Amta channel, the Damodar basin looks like a tadpole. Since the upper catchment is largely overlain by granite-gneiss which does not allow quick infiltration of water and generates huge run-off during torrential rain, the lower Damodar is proverbially flood prone. The flood enquiry committee (1944) estimated the run-off coefficient between 54 and 90% of the storm rainfall. The average slope of the river declines from 19 m/km in upper reach to 19 cm/km in the lower reach and thus cannot discharge water quickly. The tidal backflow at both the outfalls creates drainage congestion and aggravates the flood. The upper Damodar basin contains many mineral resources while the lower catchment is famous for its agricultural prosperity. These two factors were causes of drastic clearing off the natural vegetation and resulted in enormous sand deposits in the bed of Damodar. The scholars working on the complex hydro-geomorphology of this basin unanimously opined that the problem of Damodar is not so much the disposal of its surplus water but that of surplus sand. The Damodar which bestowed prosperity to Bardhaman and adjoining region was labelled as the *sorrow of Bengal* for its recurrent floods.

There was no denial of the fact that the lower Damodar area has been historically flood-prone. The greatest recorded flood was to the order of 6,50,000 cusec and experienced in 1913 and 1935. The flood 1943 was of much lower intensity and carried 3,50,000 cusec of water. But even with this lower intensity, the river breached its left embankment near Amirpur downstream of Bardhaman. The countryside was marooned under two-metre depth of water, and the railway roads and telecommunication system were disrupted (Mookerjea 1992). This deepened the anxiety of the British Government as the logistic support to their army involved in the Second World War was interrupted. The Flood Enquiry Committee was constituted in 1944 under the chairmanship of Uday Chand Mahtab, the *Maharaja* of Bardhaman. This paved the way to the constitution of the Damodar Valley Corporation under the light of Tennessee Valley Authority. Prof. Meghnad Saha and Sri K. S. Ray (1944) wrote:

> it is possible to treat the Damodar river basin, at no great cost, to full measures of planned reclamation, and thus convert a destructive river system into a beneficial agency, producing large amount of electrical power, ensuring water for irrigation and flushing the river basins throughout the year, removing the eternal menace to rail and road communication, and guaranteeing public health. Nature, vested interests and thoughtless managements made a once prosperous valley a wilderness but Nature, Man and Science can again make it a smiling garden.

The flood enquiry committee made its final recommendation on the 10 March 1944. The committee explored rainfall in the Damodar basin during preceding thirty-three years and found that the average annual rainfall in the Damodar Valley was 1181 mm. The highest annual rainfall was recorded in 1917 amounting to 1709 mm. while in the year 1915 lowest annual rainfall (779 mm) was recorded. This wide variation in annual rainfall led to hydrological two extremes—flood and drought.

In accordance with the variation in both monsoon and non-monsoon rainfall discharge in the Damodar varied from one year to other. Though the peak discharge of 6,50,000 was recorded at Rhondia in August 1913 and in August 1935, the average monsoon discharge in the latter year (in 1935) was lowest ever during the period taken into account. This means that an unusually concentrated storm rainfall caused disastrous flood in 1935. This kind of event was experienced many times in the Damodar basin. Tables 10.1, 10.2 and 10.3 describe variable annual rainfall in the Damodar valley and the

Table 10.1 Average annual rainfall of the Damodar valley (1912–1944)

Year	Average rainfall (mm)	Year	Average rainfall (mm)	Year	Average rainfall (mm)
1912	813	1923	1238	1934	1081
1913	1320	1924	1285	1935	874
1914	887	1925	1086	1936	1430
1915	779	1926	1190	1937	1193
1916	1258	1927	984	1938	969
1917	1709	1928	1184	1939	1341
1918	1004	1929	1344	1940	973
1919	1489	1930	1322	1941	1354
1920	1205	1931	1075	1942	1426
1921	1076	1932	990	1943	1351
1922	1403	1933	1177	1944	1216

Source Report of the Damodar Flood Enquiry Committee (1944)

peak discharges which flowed through the river. These data were considered to plan the Damodar Valley Project.

In view of observations made by the committee, a *preliminary memorandum on the unified development of the Damodar (PMUD)* river was prepared.

10.3.2 PMUD Summary and Observations

The principal objective of PMUD was to prepare a coordinated plan of development for the Damodar basin. The memorandum, at the outset, discussed physical, industrial and social status of the basin

Table 10.2 Average discharge and run-off of the Damodar River at Rhondia (1933 to 1944–45)

Year	Monsoon		Non-monsoon	
	Average discharge in cusec	Average run-off in inches	Average discharge in cusec	Average run-off in inches
1933–34	25,600	19.00	1020	1.05
1934–35	14,600	10.80	289	0.29
1935–36	13,600	10.10	321	0.33
1936–37	27,400	20.30	2970	3.06
1937–38	20,400	15.20	473	0.49
1938–39	18,900	14.00	335	0.34
1939–40	31,700	23.50	2050	2.11
1940–41	18,200	13.50	451	0.46
1941–42	31,200	23.20	1613	1.66
1942–43	34,900	25.90	NA	NA
1943–44	36,500	27.10	NA	NA
1944–45	30,600	22.70	NA	NA
Average	25,300	18.70	1060	1.09

Table 10.3 Peak discharge in cusec of Damodar at Rhondia (1823–1959)

Years	Discharge	Years	Discharge	Years	Discharge	Years	Discharge
1823	650,000	1911	220,000	1938	110,000	1949	272,000
1840	640,000	1913	650,000	1939	282,000	1950	338,000
1855	350,000	1914	210,000	1940	310,000	1951	389,000
1860	350,000	1915	160,000	1941	625,000	1952	181,000
1864	340,000	1916	395,000	1942	375,000	1953	293,000
1865	450,000	1917	395,000	1943	296,000	1954	262,000
1866	420,000	1933	226,000	1944	190,000	1955	100,000
1877	500,000	1934	170,000	1945	121,000	1956	303,000
1878	250,000	1935	650,000	1946	314,000	1957	201,000
1902	220,000	1936	250,000	1947	260,000	1958	665,000
1907	240,000	1937	210,000	1948	230,000	1959	800,000

Data Source Saha (2008)

which was the home of about half a million population. The population, then living in the basin were largely farmers but considering the existence the Gondwana coal seam in the subsurface, the flood enquiry committee realized the potential scope for industrial development in the area. The following issues were taken into consideration.

1. Flood Control:
   In view of the experiences of earlier floods, the top priority was assigned on flood control. The plan of development took into account the peak discharge of one million cusec at Rhondia that could be generated out of 135 inches (343 mm) of storm rainfall.
2. Storage Reservoirs:
   There was a plan to build a system of eight dams and a barrage. The topography of the basin made it impossible to locate any dam below Dissergarh where Barakar joins the Damodar. The aggregate storage of eight dams was estimated as 4,700,000 acre-feet or 5797 million m^3. That storage was expected to reduce the maximum design flood to a flow of about 250,000 cusec at Rhondia. The experts thought that the maximum possible flood from the uncontrolled catchment of the river below the lowest practicable dam site could, in any case, amount to some 20,000 cusec and as such committee found no justification of increasing the flood storage volume.
3. Irrigation:
   The area served by the Rhondia weir–Jujuti sluice–Kanchan Nagar anicut–Eden canal and Jamalpur regulator system was 186,000 acres prior to construction of reservoirs but that area was partially irrigable due to the dwindling lean season's flow in the Damodar. The proposed plan was expected to ensure perennial irrigation in 760,000 acres including the 186,000 acres that were earlier partially irrigated.
4. Power:
   The memorandum was optimistic for combining hydroelectric plants with the generating capacity of 200,000 kW along with some 150,000 kW of thermal power generating capacity in large modern units. Taking into account the transmission–distribution loss, the system was expected to generate 300,000 kW, making available about 1420 million kW hours annually.
5. Navigation:
   The experts of PMUD were speculative about the improvement of navigation channel and noted that *taming of this river with the concomitant of a substantial minimum flow would, however, bring the establishment of a navigable waterway within the realms of practical consideration* (Mathews 1947).

The Central Technical Power Board (CTPB) sought the opinion of W. L. Voorduin, an expert from Tennessee Valley Authority. After detailed investigation, Voorduin proposed the construction of (a) seven dams at Tilaiya, Deolbari, Maithon, Aiyar, Sonalpur (near Panchet Hill), Konar and Bokaro. He added that the proposed dams would be capable of accommodating a million cusec of flood water and keep the monsoonal discharge of the Damodar within 0.25-million cusec at Rhondia; Voorduin further proposed the establishment of (b) hydro-power generating station at each dam; (c) a thermal power station at Bokaro; (d) a barrage across the Damodar to ensure perennial irrigation to agricultural fields of Bardhaman, Hugli, Haora and Bankura; (e) power transmission lines; (f) and a low diversion dam and power canal at Bermo where the river debouches sharply.

Voorduin suggested the dams should be located in such way that the uncontrolled area downstream of dams should not be of such size that it could produce a flood greater than the controlled flood from the upstream area, if a maximum storm was centred over the uncontrolled area. But the topography below Dishergarh where the Damodar and the Barakar meet was not favourable for building any dam which could accommodate substantial discharge. The basin area below Dissergarh covers about 3700 km^2 which could generate substantial flow during storm rain. Voorduin added that '*shape of this area is not conducive of high flood rates because its width is small as compared to its* length and natural valley storage in the lower Damodar River would considerably reduce any flood peaks. It is believed therefore, that the peak flow of any flood does not substantially increase in magnitude below the Anderson Weir at Rhondia'.

Making of DVC Dams

Shortly after the independence, an Act (XIV of 1948) was passed in the Parliament of India to facilitate the execution of the project.

Apart from flood control, two other major objectives of the DVC project were irrigation and power generation. It was expected that project would generate 1.50 lakh kW of power and irrigate 3,29,826 ha of land through 2495-km-long irrigation canals (Fig. 10.1 and Table 10.4).

But due to fund crunch, the Government decided to phase out the proposal of Voorduin. Initially, it decided to construct four reservoirs at Tilaiya, Konar, Maithon and Panchet instead of Sonalapur. Subsequently, Bihar Government built

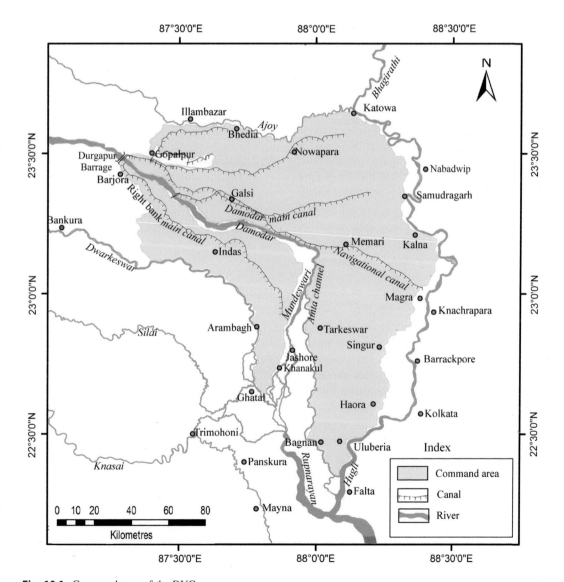

Fig. 10.1 Command area of the DVC

Table 10.4 DVC Dams at a glance

	Tilaiya	Konar	Maithon	Panchet
Commissioning	1953	1955	1957	1959
On river	Barakar	Konar	Barakar	Damodar
Length (m)	366	4535	4860	6777
Power generating capacity	4 MW	–	60 MW	80 MW
Storage capacity (mcm)				
Dead storage	75	60	207	170
Total storage	395	337	1362	1498
Allocation in different sectors (mcm)				
Irrigation & power	142	221	612	228
Flood control	178	56	543	1087
Catchment area (km^2)	984	997	6293	10966
Reservoir areas (km^2)				
At dead storage level	15.38	7.49	24.28	27.92
At maximum conservation pool	38.45	23.15	71.35	121.81
Area top of gates	74.46	27.92	107.16	153.38

Source Compiled from http://www.dvc.gov.in

a reservoir at Tenughat in lieu of Aiyar. Unlike four other reservoirs under DVC, the last one does not have any flood cushion and is a two-tier reservoir with dead and live storage capacity (Fig. 7.5). Since the Government could not acquire the total land required for the Maithon and Panchet reservoirs, the storage capacities in both cases were compromised and water levels of the reservoirs remained 1.50 and 6 m below the desired levels. However, the five reservoirs together initially accommodated 56.44% water of the total estimated capacity of seven reservoirs.

All reservoirs of the world are losing their storage capacity at the average rate of one per cent per year and so the lifespan of a reservoir cannot be longer than a century. Since the DVC reservoirs are within tropical climate region and the basin is degraded, the rate of sedimentation in reservoirs is very high. It was estimated that the rate of sedimentation in Maithon reservoir would be 0.84 × 10^6 m^3/year but the observed rate has been 7.30 × 10^6 m^3/year. The sedimentation in Panchet reservoir has been four to five times higher than the anticipated rate. The age-old mining, deforestation and expansion of agriculture have contributed substantial sediment in the river leading to the decay of the channels as well as reduction of storage capacities of reservoirs (Table 10.5). All four reservoirs have lost their storage capacity between 23 and 43% till 2016.

10.3.3 Power Generation and Irrigation

The second major objective of the Damodar Valley Corporation was to generate both hydro-power and thermal power (Table 10.6). But there was a conceptual void as it was difficult to ensure both irrigation and power generation from the same reservoir. The generation of power requires uninterrupted release of water from the reservoir to rotate the turbine; while irrigation requires the storage and transfer of the monsoon water for the non-monsoon season. Since there is no separate allocation of water in the DVC reservoir for power generation and irrigation, one of the objectives has to be sacrificed to ensure the other. It was next to impossible to serve the dual

Table 10.5 Reducing capacity of the DVC reservoirs

Reservoir/year of commissioning	Catchment area (km^2)	Initial capacity (MCM)	Rate of silting (MCM/year)	Loss of capacity till 2016	
				MCM	% Loss
Maithon (1955)	6294	1362	6.77	413	30
Panchet (1956)	10,878	1498	5.57	340	23
Tilaiya (1953)	984	395	2.81	169	43
Konar (1955)	997	337	1.75	108	32

Source Compiled and Computed from Report CWC (2001)

Table 10.6 (A) Thermal power generation by the DVC. (B) Hydro-power stations of the DVC

(A)			
Name	Location	Capacity (MW)	Commissioning
Bokaro	Dist—Bokaro State—Jharkhand	630	Between 1986 and 1993
Chandrapura	Dist—Bokaro State—Jharkhand	390	Between 1964 and 1968
Durgapur	Dist—West Bardhaman State—West Bengal	350	Between 1966 and 1982
Mejia	Dist—Bankura State—West Bengal	1340	Between 1996 and 2008
Total Thermal		**2710**	
(B)			
Name	River	Existing capacity (MW)	Commissioning
Tilaiya	Barakar	4	1953
Maithon	Barakar	60	1957–1958
Panchet	Damodar	80	1959 and upgraded in 1991
Total Hydro-power		**144**	
Grand total		**2854**	

Source http://www.dvc.gov.in

purposes from a single reservoir. It is to be noted that the total installed power generating capacity of the DVC is 2854 MW, of which hydro-power contributes only 144 MW or 5%.

The four reservoirs supply water to Durgapur barrage which was supposed to ensure irrigation in 394,000 ha of land annually through a network of 2494-km-long canals. The maximum possible target was to irrigate 393,768 ha of paddy field in rainy season, 18,450 ha of winter crop and 69,790 ha of dry season paddy (www.wbiwd.gov.in). In addition, 30,000 ha of land within controlled catchment were supposed to be irrigated by lift irrigation from 16,000 check dams. But the area actually irrigated is fell short of expectation, and farmers of DVC command area rely largely on groundwater which has gone down due to overexploitation.

10.3.4 Flood Control

While preparing the Preliminary Memorandum of Plan of Unified Development (PMUD) of the Damodar River, W. L. Voorduin noted that '*the primary consideration for a plan of development* of the Damodar Valley should be control of floods. It is further deemed desirable that insofar as possible the system of dams should be capable of producing the largest amount of power which could be made available and that the maximum use be made of the regulated flow for irrigation purposes'. But the primary objective of building DVC reservoirs was proved futile twice in 1958 when incessant rain of the 4 July and 16–17 September proved the inadequacy of the newly introduced flood management system. The Government of West Bengal formed a committee in 1959 under the chairmanship of Sardar Man Singh.

The Committee concluded in its final report—'*Before the construction of D.V.C dams, the flood peaks were high but duration was small. The construction of dams moderated the peaks but increased the duration of floods. This increase in duration has enhanced the chances of synchronization of floods from the upper and lower valleys as also from adjoining river basins*' (Gov. WB 1959). The increase in duration of flood may be attributed to the decay of distributaries as well as the expansion of the intricate network of railways and roads in the lower Damodar basin. Many people believe that the phasing out of the project reducing number of reservoirs and their total water-retaining capacity might be the cause of the menace. However, it is recorded that the discharge released from the Durgapur barrage exceeded the threshold limit 2.50-lakh cusec in seven years during the period 1956 to 2015 and the highest discharge was released in 1978 (Table 10.7).

Mr. Debes Mookerjea (1992) former General Manager and Secretary of DVC noted—'*Flood control activities have become the victim of mutual mistrust and truncated fundings. To keep down the cost of flood control, Government of West Bengal observed that they would be satisfied with the provision of adequate storage capacity for high flood of the intensity of 6.5 lakh cusec moderated to bankful capacity of the Damodar in the lower reaches which was 2.5 lakh cusec in the forties .... During last three or four decades, the bankful capacity of lower Damodar has shrinked to much less than even 100,000 cusec*'.

The altered hydraulic regime of the lower Damodar area has been one of causes of the recurrent floods. In September 1978, torrential rain in the catchment of the reservoirs compelled the DVC authority to release 160,000 cusec of water which synchronized with 220,000 cusec of water generated in uncontrolled lower catchment and the total flow of 380,000 cusec marooned about 86% area of the lower Damodar basin. On the 22 September 2000, Durgapur barrage released 230,000 cusec of water and that water rushed over extensive areas of Bardhaman, Hugli and Haora.

No one denies that the DVC was dedicated to the nation with the pious intention of service to mankind. But the gap between the irrigation potential promised and that actually achieved continues to increase with the reduction in the capacity of reservoirs. The transmission–distribution loss of water is so much that areas lying in the lower reach of the command area are not served. The farmers continuously exploit groundwater table which tends to deplete. The concept of controlling flood by storage of water in reservoirs remains elusive. Thus, the experiences of the last six decades were often bleak but lessons were important.

10.3.5 DVC: An Overview

The Damodar Valley Corporation was a very optimistic project under the plan of remaking the nation during post-independence era. But the initial objectives of the British rulers were to tame the river which breached the communication lines connecting Kolkata with north India during the devastating floods of 1943. They aimed at protecting the army base at Panagarh, Bardhaman town, Grand Trunk road and the Railway to ensure the logistic support to the British army involved in Second World War. A look at the discharge data of the river Damodar at Rhondia for the period prior to construction of DVC reservoirs (1827–1959) reveals that the

10.3 Post-Colonial Period

Table 10.7 Peak discharge (cusec) of Damodar at Rhondia (1956–2015)

Years	Discharge	Year	Discharge	Years	Discharge	Years	Discharge	Years	Discharge
1956	307,100	1968	161,511	1980	162,395	1992	36,550	2004	67,700
1957	201,000	1969	53,675	1981	55,097	1993	136,466	2005	37,375
1958	305,150	1970	138,139	1982	21,950	1994	117,459	2006	271,675
1959	428,525	1971	222,368	1983	78,250	1995	304,800	2007	295,700
1960	115,625	1972	38,550	1984	182,515	1996	123,795	2008	94,100
1961	189,805	1973	199,731	1985	122,121	1997	NA	2009	311,600
1962	NA	1974	104,578	1986	125,731	1998	144,945	2010	86,000
1963	132,450	1975	154,065	1987	188,525	1999	216,200	2011	143,028
1964	62,450	1976	199,495	1988	50,575	2000	223,292	2012	85,723
1965	89,575	1977	141,120	1989	77,175	2001	65,410	2013	163,250
1966	8100	1978	379,326	1990	117,975	2002	59,099	2014	74,000
1967	135,950	1979	36,225	1991	86,550	2003	80,993	2015	128,250

Data Source Saha (2008) and Irrigation department Government of West Bengal

flood discharges exceeding six lakh cusec have occurred in the years 1823, 1840, 1913, 1935, 1941, 1958 and 1959. The DVC reservoir since 1959 regulated the peak discharge, and the flood of such high magnitude was never experienced thereafter. But it should be noted that the basic objective of DVC reservoirs to regulate the flow within 2.5-lakh cusec at Rhondia was not fulfilled in the years like 1956, 1958, 1959, 1978, 1995, 2006 and 2007, and the flood continues to imperil the area as usual.

The claim of DVC creating irrigation potential of 394,000 ha of land annually by the stored water in four reservoirs seems untenable when crop water demands are taken into account. Considering the irrigation demand of Boro (1500 mm), Amon (1000 mm) Rabi (400 mm) and the potential area to be cultivated for these three crops, the total water demand amounts to 3958 mcm, which is more than the combined storage capacity of the four reservoirs.

The DVC authority often takes a calculated risk of storing the water even to the height of highest pond level in the month of July and August with the pious intention of ensuring irrigation during Rabi and Boro season. But they are supposed to keep the uppermost tier of the reservoir vacant to accommodate the water generated out of late monsoon cloudburst which has earlier caused disastrous floods in the lower Damodar area. Since late monsoon cloudburst is an uncertain meteorological phenomenon, most of the time the dam managers may not be in trouble by maintaining the reservoir in its full capacity; but in years of excessive precipitation in the month of September, (as was in the case of 1978 and 2000), they were forced to open the gates of reservoirs causing dam-induced floods in its lower reaches.

The gap between the target of irrigation and actually irrigated area has widened considerably, and two other causes of this menace are transmission–distribution loss of water and continued sedimentation in reservoir. The Ministry of Water Resource, the Government of India, has stated that the average water efficiency of all large dams–canals network in India is less than 40% and the DVC is not an exception. The demand of water for irrigation has increased many times with the introduction multiple cropping. The demand is so high that no ecological flow is released from the Durgapur barrage during lean months.

10.3.6 Mayurakshi Project

The district of Birbhum is identified as historically drought-prone though it receives an annual rainfall of about 1527 mm. The water scarcity in this district is largely due to its undulating terrain capped with laterite which does not allow easy infiltration and generates huge run-off. The five minor tributaries, namely Matihara, Dhobbi, Pussaro, Bhamri and Tepra originating from uplands of Dumka are combined to form Mayurakshi, literally meaning peacock's eye. The river owes its name from the silt-free and transparent water that flows during the lean months. The Mayurakshi project was initiated in 1954–55, and it includes following components described in Tables 10.8 and 10.9.

Thus, the Mayurakshi irrigation project includes one reservoir, four barrages and a weir. The 361-km-long main canal and 1261-km-long distributary canals have been excavated under this project. The irrigation potential created under this project covers about 0.247 million ha, and this is shared by Birbhum (72%), Murshidabad (22%) and Bardhaman (6%) districts. After six decades of commissioning of the Massanjore reservoir, its storage capacity has appreciably reduced due to sedimentation. According to the Central Water Commission, the rate of sedimentation in Massanjore reservoir is 31.90 lakh m^3/year (Fig. 10.2).

This means that storage capacity in 2014 stands 431.50 million m^3 which cannot adequately serve the targeted irrigable area under Rabi crop. It is difficult to keep the reservoir bank-full at the beginning of Rabi season, as the supplementary irrigation for Kharif and huge demand for *Boro* (dry variety paddy) demands withdrawal of principal share of water from the reservoir. The gap is further widened when approximately two cumec (67 cusec) of water is diverted from Tilpara barrage to the Bakreshwar thermal power plant.

It is important to note that unlike DVC reservoirs, Massanjore reservoir can store water in two tiers, called the dead storage and live storage, respectively. It does not have any space as flood cushion. But common people living in lower catchment expect the reservoir to regulate flood which is beyond its capacity. The hard reality is that the lifespan of the Mayurakshi project is close to its end. The decentralized rainwater harvesting seems to be the only option to meet the ever-increasing demand for irrigation.

10.3.6.1 Kanshabati Project

The Kanshabati project was contemplated to ensure irrigation in relatively water-short areas of Bankura and undivided Medinipur. The construction of Kanshabati barrage was started at Mukutmanipur in 1959. The barrage was placed at the confluence of Kanshabati and Kumari rivers. Two other barrages, namely Tarafeni and Bhairabbanki, were built on two right bank tributaries of Kanshabati, and these two small reservoirs replenish the main reservoir.

A 28-km-long left bank canal connects the Kanshabati reservoir with Silai where another barrage was built. Thus, the project as a whole comprises of one major reservoir on main river, and three barrages on Silai, Bhairabbanki and Tarafeni. The 805-km-long main canal with 2425 km-long-distributary canals was excavated to irrigate 340,890 ha land for Kharif cultivation and 60,718 ha for Rabi cultivation (Fig. 10.3). The technical detail of the Kanshabati project is described in Tables 10.10 and 10.11.

Table 10.8 Irrigation potential created (in ha)

Districts	Kharif crop	Rabi crop	Total
Birbhum	160,931	16,210	177,141
Murshidabad	49,797	4033	53,830
Bardhaman	15,993	–	15,933
Total	226,721	20,243	246,964

10.3 Post-Colonial Period

Table 10.9 Components of the Mayurakshi project

1. Dam cum reservoir at Massanjore		
A.	Area of the reservoir	10.42 km^2
B.	Storage capacity	616.5 million m^3
C.	Controlled catchment above the dam	1860 km^2
D.	Estimated peak discharge	300,000 cusec
2. Tilpara barrage		
A.	Length	307 m
B.	No of bays	15
C.	Width of each bay	18.18 m
D.	No of head regulator	2
E.	Catchment	3209 km^2
F.	Estimated peak discharge	200,000 cusec
3. Bakreshwar weir		
A.	Length	94 m
B.	Catchment	126 km^2
C.	Head regulator	1
4. Kopai barrage		
A.	Length	65.45 m
B.	Catchment	220 km^2
C.	Head regulator	1
5. Dwaraka barrage		
A.	Length	108 m
B.	Catchment	303 km^2
C.	Regulator	1
6. Brahmani barrage		
A.	Length	142 m
B.	Catchment	671 km^2
C.	Regulator	1

Like all other large dam–canal projects, there is a wide gap between the target and achievement. This is largely due to the misconception of the planners at the formative stage of the project. The potential created for the cultivation of Rabi crop needs 607.18 million m^3 of water if the average crop water demand is 1000 mm. But the reduced capacity of reservoir due to sedimentation, the evaporation as well as transmission–distribution loss of water during long distance transfer of water is responsible to keep the irrigated area far short of the target.

10.3.7 The Farakka Barrage Project

The emergence of the port of Kolkata has a long colonial history. The European merchants first visited Bengal during the early sixteenth century. Satgaon was then the most important trading centre on the bank of the Saraswati, a distributary of the Bhagirathi-Hugli River. But it was extremely difficult to approach the port from the sea because of the decaying status of the Saraswati. The large ships had to anchor at Betor near present Shibpur of Haora district and only

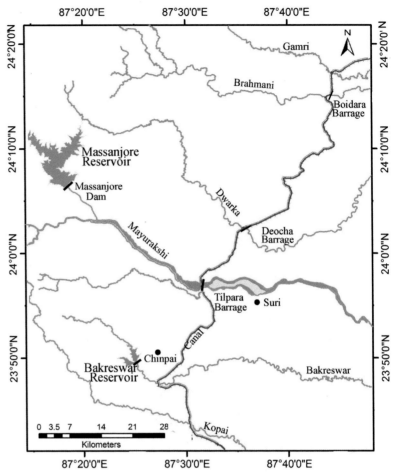

Fig. 10.2 Map showing the layout of the Mayurakshi project

smaller wooden boats, *Bazras*, were able to reach Satgaon during the high tide (Mukherjee 1938).

The British merchants started to operate their business from Hugli in 1651 and that continued till the seventeenth century (Nair 1995). The conflict with the Mughals compelled the British to move southwards in search of a new port. The Kolkata gradually emerged as a port since 1690 when Job Charnock was given right of trading by the Mughal Emperor Aurangzeb. The trading operation got momentum in the eighteenth century when light ships had been sailing up and down the Hugli River (Ghosh 1972). However, navigation was a difficult task due to inadequate depth of river. The hinterland of the port of Kolkata gradually extended from Assam to Uttar Pradesh. The navigation became further difficult when large vessel appeared in the scene. It was only during high tide seagoing vessels could approach the port of Kolkata. The dredging of the navigation channel to render safe passage to ship started since 1820 but such measure could not ensure permanent solution (Sanyal and Chakraborty 1995).

The Bhagirathi-Hugli River was gateway of Portuguese, French, Danish and British merchants in the sixteenth and seventeenth centuries and the latter ultimately took the administrative control. The Kolkata emerged as port-industrial city from three sleepy villages. The riverfront changed from agricultural to industrial landscape. The congregation of many workers from neighbouring States increased the density of population and finally large-scale migration across the international border of divided Bengal in 1947 brought a dramatic change in geographical

10.3 Post-Colonial Period

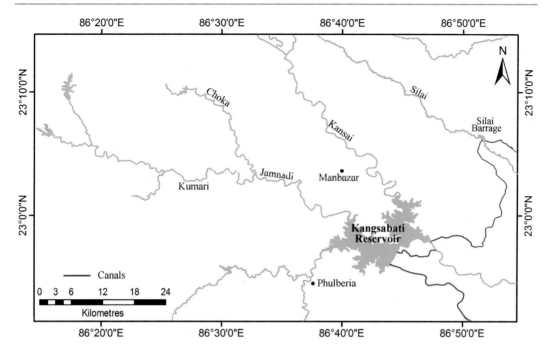

Fig. 10.3 Kanshabati project

Table 10.10 Technical details of the reservoir at Mukutmanipur

A.	Length of dam	11.27 km
B.	Total storage capacity	1056 million m^3
C.	Live storage	900 million m^3
D.	Main canal	805 km
E.	Branch canals	2425 km
F.	Controlled catchment	3626 km^2
G.	Uncontrolled catchment	8220 km^2

Other barrages A. Silai, B. Bhairabbanki, C. Tarafeni

Table 10.11 Irrigation potential (in ha) created by the Kansai project

District	Monsoon paddy	Winter crop
Hugli	19,167	4039
Bankura	153,723	24,291
Medinipur	168,000	32,388
Total	340,890	60,718

scenario. What had been an anchorage gradually developed as an international port.

Decay of the River

The difficulties in navigation in the Hugli estuary were a matter of deep anxiety for the British, as colonial trade was intimately related to navigation in this part. The decay of the Bhagirathi-Hugli River, however, started long before the establishment of the Kolkata port. In 1853, Bengal Chamber of Commerce proposed the establishment of a substitute port on the bank of Matla River, some 40 km to the south of

Kolkata. The port was named after Lord Canning. The newly built port became moribund within a decade and was officially closed in 1871 (GoI 1865). It was then obligatory for the British to explore the ways and means to deepen the channel leading to Kolkata port. The engineers first tried to cut open the off-take point of the Bhagirathi to induce water from the Ganga. But such attempts proved futile as the water level of Ganga fell five metres below the bed level of Bhagirathi during lean months. The off-take ceaselessly migrated from one point to other between 1781 and 1925 (Mitra 1953) and the process continued till 1975 when the inlet between the Ganga and the Bhagirathi at Jangipur was closed to restrict the reversal of canal induced water.

Project of Resuscitation

In 1853, Sir Arthur Cotton recommended for diversion of the water from the Ganga to the Bhagirathi to improve navigational facility. The issue was discussed at length and probable site for the barrage was explored. The project of resuscitating the port of Kolkata was shelved till the independence when the Government in independent India decided to act upon the proposal. The Government of India sought the opinion of W. Hensen, an eminent expert from US hydraulic study department in 1957. After detailed survey and research, he opined that

> '*The best and only technical solution of the problem is the construction of a barrage across Ganga at Farakka with which the upland discharge into the Bhagirathi-Hooghly can be regulated as planned, and with which the long term deterioration in the Bhagirthi-Hooghly can be stopped and possibly converted into a gradual improvement. With a controlled upland discharge a prolongation of freshet period will be obtained, and the sudden freshet peaks which will cause heavy sand movement and bank erosion will be flattened*' (GoI 1975). The 2.62 km long Farakka barrage was built across the Ganga between 1962 and 1971. The 38 km long Ganga-Bhagirathi link was excavated between 1971 and 1975. Table 10.12 and Fig. 6.3 describe the different components and layout of the Farakka barrage project.

It has been subsequently observed that the sediment dynamics in the Hugli estuary is governed by a complex fluvial system which cannot be altered by inducing 40,000 cusec (1132 cumec) of flow. In this estuary, the tidally induced flow dominates over the upstream supply of water. It is estimated that the peak flood and ebb discharges vary to the order of 2.6×10^5 cubic metre per second and 1.09×10^5 cubic metre per second respectively (McDowel and O'Connare 1977). This means that the volume of water entering through the estuary during high tide is about 230 times larger than the water induced from the Farakka. The south-flowing freshwater in the Hugli even during the rainy season is not strong enough to reverse the impact of silt-laden tidal inflow. Thus, the sediment management appeared to be a challenging task.

Increased Quantum of Dredging

The commissioning of the Farakka barrage was expected to improve status of navigation channel approaching the port of Kolkata. The water diverted from the Ganga to the Bhagirathi was expected to reduce the sedimentation in the estuary and ensure better draught for the ships (Fig. 10.4). But the vision was frustrated and sedimentation remained an unsolved problem. The Kolkata Port Trust reported the quantum of dredged material was 6.40 mcm prior to 1975 but subsequently increased to 13.24 mcm (Sanyal and Chakraborty 1995). In 1999–2003, the quantum further increased to 21.18 mcm. The tide-velocity asymmetry is one of the major causes of sedimentation. The tidal bore moves fast with silt-laden water and tends to be deflected towards right-hand side due to Coriolis force and the comparatively slower ebb flow facilitates sedimentation on the estuary.The tidal swelling of water may be from 1.51 to 6.60 m at sea front and ranges from 1.47 to 7.35 m. at Diamond Harbour (KPT 2016). It appears to be a challenging task to flush to sediment load from the estuary to the deeper sea and exploring a better location further south may be a better option. It is important to

Table 10.12 Technical details of the Farakka barrage Project

A. *Farakka barrage*	
Length	2.62 km
Number of bays	109
Width of each bay	18.30 m
Lowest bed level	10.30 m above msl
Pond level	21.90 m above msl
Crest level of spillway	15.80 m above msl
B. *Head regulator*	
Pond level	21.90 m above msl
Full supply level at land	1133 cumec
Waterway	11 bays of 12.20 m each
Crest level	18.10 m. above msl
C. *Feeder canal*	
Length	38.30 km
Design discharge	1133 cumec
Bed width	150.80 m
Full supply depth	6.10 m
D. *Jangipur barrage*	
Length	212.70 m
Number of bays	15
Width of each bay	12.2 m
Crest level	12.80 m above msl

Source Basu (1982)

realize that port of Kolkata, being located about 150 km inland from the sea has been suffering due to the complex fluvio-marine processes operating in the Hugli estuary. The situation became more complex when the attempts were made to make channel navigable for large vessels which require deeper draught to ply. The induced flow from the Farakka barrage has rejuvenated the non-tidal part of the Bhagirathi River but had no appreciable impact on the estuarine reach. So the construction of a deep-sea port seems to be only viable option.

10.4 The Teesta Barrage Project

The Teesta Barrage Project was conceived by the Irrigation and Waterways Department, Government of West Bengal, in 1976, with the vision of irrigating 9.22 lakh ha of land (Fig. 10.5). The area proposed to be irrigated under the first substage is 3.42 lakh ha in the six districts of West Bengal. In addition to the objective of irrigation, there was a target of generating 1000 MW hydroelectricity. The onus of supplying domestic water (26 MLD) to Siliguri Municipality was assigned to the TBP subsequently. But the project which was commissioned in December 1997 has been able to irrigate 54,171 ha till 31 March 2015. It is agreed in official level further expansion of command area is hardly possible due to dwindling discharge in the Teesta. The proposal of generating hydro-power has also been frustrated. It was possible to generate 67 MW power at canal fall structures only during rainy season. Notably, water requirement for running at least one unit to generate 22.5 MW power is 70 cumec.

Seven hydro-power stations (4 in Sikkim and 3 in West Bengal) on Teesta and its tributaries

Fig. 10.4 Hugli estuary: satellite image

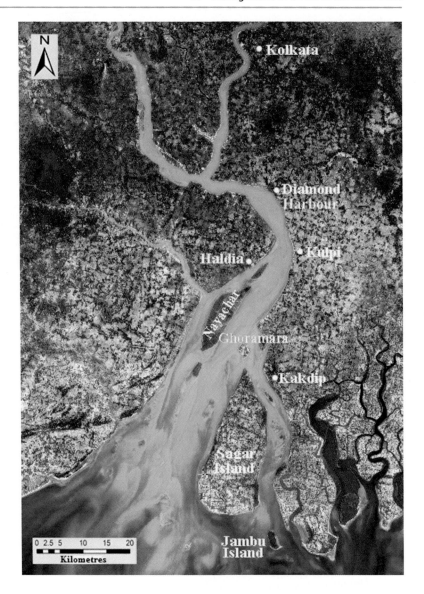

have been built so far. Five more are in pipeline. Though these are 'run of the river' schemes, still they have huge impact on stream hydrology. Water stored in these dams for 16–20 h a day during non-monsoon months and released during remaining 8–4 h to facilitate turbines to generate power. This leads to fluctuation of flow varying from 0 to 650 cumec. The downstream impacts are the total desiccation of river in 16-km-stretch between the Teesta low dam stage IV to Gazoldoba barrage affecting flora and fauna and extreme difficulty in operation of barrage gate and head regulator. The barrage on Teesta was built at Gazoldoba of Jalpaiguri district. The barrage authority needs to maintain the pond level at 114.30 m to be able to divert water through canals. If the pond level is to continuously maintained, then the total outflow through the barrage and the canals should be the same as the inflow from upstream.

The project was originally designed to connect Teesta to both Mahananda and Jaldhaka by right and left bank canals, respectively. The Teesta–Mahananda Link Canal (TMLC), which

10.4 The Teesta Barrage Project

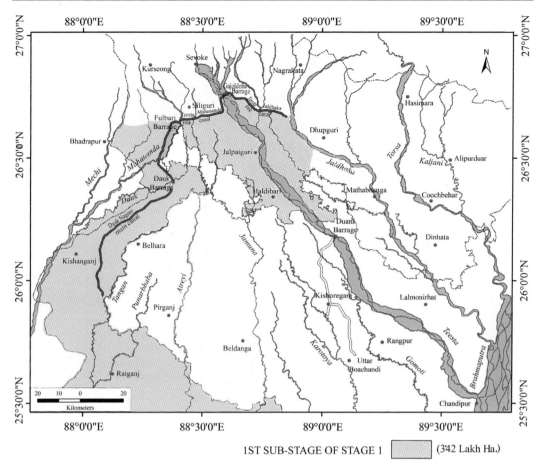

Fig. 10.5 Proposed command area under Teesta Barrage Project

takes off from the right bank of the Teesta just upstream of the barrage, with a design discharge of 415 cumec, became functional from December 1997. The Teesta–Jaldhaka Main Canal (TJMC), which takes off from the left bank with a carrying capacity of 105 cumec, has become functional from January 2012. The irrigation engineers planned to expand the command area in phases, and the project was divided into three substages (Tables 10.13 and 10.14). But now the

Table 10.13 Components of the Teesta Barrage Project

1.	Barrage at Gazoldoba
2.	Pick up barrages on: a. at Fulbari on Mahananda River b. on Dauk River
3.	Right bank irrigation canals: a. Teesta–Mahananda link canal (25.75 km) b. Mahananda main canal (32.22 km) c. Nagar–Tangon main canal (42.20 km)
4.	Left bank irrigation canal: a. Teesta–Jaldhaka main canal (30.31 km)

Table 10.14 Proposed irrigable area in TBP (in thousand hectares)

District	Stage-I		Stage-II	Stage-III	Total	Remarks
	1st substage	2nd substage				
Koch Bihar	20	61	142		223	Augmentation of Karotowa Talma Irrigation Scheme for 5600 ha in Jalpaiguri district included in the first substage
Darjeeling	17				17	
Jalpaiguri	62		81		143	
Uttar Dinajpur	194				194	
Dakshin Dinajpur	10	121		65	196	
Maldah	39	22		88	149	
Total	342	204	223	153	922	

Source Irrigation and Waterways Department, Government of West Bengal (1996)

Table 10.15 Gap between target and achievement

Canal	Designed length of canal(km)		Presently operational (km)		Designed CCA (ha)	CCA presently irrigated (ha)
	Main	Distributary	Main	Distributary		
TJMC	30.31	242.22	30.31	34.77	38,000	4461
TMLC & KTIS	25.75 25.11	276.06 44.03	25.75 25.11	158.74 44.03	47,352	36,040
Mahananda main canal	32.22	249.69	32.22	86.03	45,871	9785
Dauk-Nagar main canal	80.62	394.41	70.20	31.08	94,733	3885
Nagar–Tangon main canal	42.20	664.79	0.00	0.00	116,114	0.00
Total (substage I)	236.21	1871.20	183.59	354.65	342,070	54,141

Source Records of Irrigation department (TJMC—Teesta–Jaldhaka main canal; TMLC—Teesta–Mahananda link canal; KTIS—Karatoya-Talma irrigation scheme)

project seems to be too unrealistic as water presently flowing through the Teesta is not enough even to serve the command area planned for first substage, which had a target to irrigate 3.42 thousand hectares. Unfortunately, less than 16% of the targeted irrigation area has hitherto been achieved. A critical look at the project reveals the wide gap between target and achievement (Table 10.15).

It appears from Table 8.17 that while there was a target to irrigate 342,070 ha of land under substage I, only 54,141 ha were irrigated. This is due to diminishing flow in the Teesta and creation an unrealistic command area.

Table 10.16 transpires that irrigation command area targeted for the first substage appears grossly impossible to achieve; it is striking to note that while average flow of Teesta at barrage in the month of March is 120 cumec, the demand of water in the first week of the same month is 239 cumec. The real picture is more worst as flow fluctuates during night-time and daytime due to operation of hydro-power projects, especially the stage IV low dam located about 16 km

Table 10.16 Water-demand for first substage and supply (target CCA 342,000 ha)

Months	Crops presently cultivated	Total water requirement at canal main heads for all crops (cumec)	Estimated monthly flow at (cumec)
December	Boro, wheat, winter, vegetable, oilseeds	57 (1–10 Dec.) 77 (11–20 Dec.) 79 (21–31 Dec.)	149
January	Do	90 (1–10 Jan.) 90 (11–20 Jan.) 80 (21–31 Dec.)	116
February	Same as considered in December with addition of maize	105 (1–10 Feb.) 121 (11–20 Feb.) 77 (21–28 Feb.)	101
March	Boro, jute, wheat, summer vegetable, maize, oilseed	239 (1–10 March) 237 (11–20 March) 102 (21–31 March)	120
April	Boro, jute, wheat, summer vegetable, maize, oilseed	125 (1–10 April). 135 (11–20 April) 140 (21–30 April)	175

Source Unpublished records of Irrigation Department, Government of West Bengal

upstream of the Gajoldoba barrage. It is admitted that Teesta barrage can serve to ensure supplementary irrigation during the monsoon and at best serve 54,141 ha during lean months.

10.5 The Teesta Barrage Project in Bangladesh

Following the Gazoldoba barrage in Indian, Bangladesh constructed a barrage at Duani (Lalmonirhat district) in 1990. In this case also, irrigation was dependent on flowing water in the river, as no reservoir was built to store and transfer monsoon water for non-monsoon season. The project in Bangladesh was designed to irrigate 111,406 ha in phase I which was completed in June 1998. The project proponents hope to expand the irrigable area to 7,49,000 ha in phase II (Islam et al. 2004).

Similar to the case of the Indian TBP, the planners of the Bangladesh TBP hardly took into account available water at barrage site. Consequently, there remained a wide gap between potential command area and the area actually irrigated. The project is only able to help in supplementary irrigation to *Kharif* crop during mid-monsoon breaks. In fact, the gap between expectations and achievement at the Bangladesh TBP came to be attributed entirely to water diversion at the Indian TBP, adding to the stress in bilateral relationship.

The recent hydro-politics and negotiation between India and Bangladesh on the sharing of Teesta water got much focus in the media. Since Bangladesh built a barrage across Teesta, the unrealistic optimism in both India and Bangladesh over the irrigation potential of two projects added has stressed the bilateral relationship. The issues are discussed at length in Chap. 11.

References

Basu SK (1982) A geotechnical assessment of the Farakka barrage project, Murshidabad and Maldah districts, West Bengal. Bulletin, Geological Survey of India, 47:2–3
Biswas A (1981) The decay of irrigation and cropping in West Bengal (1850-1925). Cressida Transaction, 1(1)
Chapman GP, Rudra K (2007) Water as foe, water as friend: lesson from Bengal's millenium flood. J S Asian Dev 2(1):19–49
Crooke W (1925) Tavernier's travels in India, vol I
D'Souza R (2006) Drowned and dammed/colonial capital and flood control in Eastern India. Oxford, New Delhi

Ghosh HP (1972) History, Development and problems of dredging in the Hooghly river. In: Bagchi K (ed) The Bhagirathi Hooghly basin. University of Calcutta, pp 174–189

Gladwin F (1800) Ayeen Akbery, or Institute of emperor Akber, II, London

GoI (1865) Papers relating to the formation of port Canning on the river Matla. Bengal Printing Company Ltd, Calcutta

GoI (1975) Report on the preservation of the port of Calcutta. Ministry of Irrigation and Power, New Delhi

GoI (2001) Compendium on silting of reservoirs in India. Central Water Commission, Govt. of India

Govt. of West Bengal (1959) Final report of the West Bengal Flood Enquiry Committee

Govt. of West Bengal (1996) A note on the Teesta barrage project, irrigation and waterways department (Unpublished)

Inglis WA (1909) The canals and flood banks of Bengal, Bengal Secretariat press, Reprinted in Rivers of Bengal, V (I), 62–83, West Bengal district Gazetteers (2002)

Islam N, Azam A, Islam QR (2004) Teesta river water sharing: a case study in Teesta barrage project. Second Regional Conference of the International Commission on irrigation and drainage. Available online (http://watertech.cn/english/islam.pdf)

Kolkata Port Trust (2016) Tide tables for Hugli river, Kolkata

Mahalanobis PC (1927) Rainfall and floods in North Bengal, 1870–1922. Department of Irrigation, Bengal Government

Majumdar SC (1941) Rivers of the Bengal Delta, Calcutta

Mathews HM (1947) Preliminary memorandum on unified development of Damodar river/summary and observations. Reprinted in Evolution of the Grand Design, DVC, Kolkata

Mcdowell DM, O' Connor BA (1977) Hydraulic behaviour of estuaries. Macmillan Press, Cambridge

Milliman JD, Meade RH (1983) World-wide delivery of river sediment to the oceans. J Geol 91(1):1–21

Mitra A (1953) History of the mouth of the Bhagirathi river 1781–1925. Selection of records of the Government of Bengal relating to Nadia rivers (from 1848–1926), Reprinted in District Census Handbooks/Murshidabad 1961, Appendix vi, pp. 111–133

Mookerjea D (1992) DVC: a response to the challenge in Damodar valley. Evolution of the Grand Design: 115–132

Mukherjee RK (1938) The changing face of Bengal. University of Calcutta

Nair PT (1995) Early days of colonialism and the port. In: Chakraborty SC (ed) Port of Kolkata, commemorative volume. KPT, Kolkata

Ray PC (1932) Life and experience of a Bengali chemist, II, Reprinted by Asiatic Society (1996), Kolkata

Saha M (2008) Rarh Banglar Duranto Nadi Damodar, Laser Art, Sreerampur, Hooghly

Saha MN (1933) Need for a hydraulic research laboratory. Reprinted (1987) in Collected works of Meghnad Saha. Orient Longman, Calcutta

Saha MN, Ray KS (1944) Planning for Damodar Valley. Science and Culture 10(20). Reprinted (1987) in Collected works of Meghnad Saha. Orient Longman, Calcutta

Samanta A (2002) Malarial fever in Colonial Bengal/ 1820-1939/Social history of an epidemic. Firma KLM Pvt. Ltd., Kolkata

Sanyal T, Chakraborty AK (1995) Farakka barrage project: promises and achievements. In: Chakraborty SC (ed) 125 years of Kolkata Port trust, Commemorative volume, pp 55–58

Willcocks W (1930) Ancient system of irrigation in Bengal, University of Calcutta

Conflicts Over Sharing the Waters of Transboundary Rivers

11

Abstract

There are 54 rivers which cross the Indo-Bangladesh border, and the issue of sharing water has become increasingly important in hydro-politics of this subcontinent. The Indo-Bangladesh conflict over sharing of the Ganga water started in the 1970s when India built a barrage at Farakka to induce water towards the port of Kolkata. In 1996, both the countries agreed to share the water based on the available flow. Subsequently, the conflict over sharing of the Teesta water emerged as a new issue of the hydro-politics. In the 1990s, both India and Bangladesh embarked on water diversion projects on the Teesta River, through networks of canals and barrages built at Gajaldoba (India) and Duani (Bangladesh). Both the projects were created with unrealistic expectations, and they were inevitably faced with water shortage. The mutually acceptable solution lies in exploring a rational meeting point between the volume of water that may be withdrawn from the rivers and the flow to be allowed in the rivers to sustain the ecosystem.

The flow of a river is governed by the slope of land and gravitational pull. The international boundaries in many parts of the world were drawn on different considerations but paid no heed to watercourses and their basins and thus subsequently led to the conflicts over sharing the water. A transboundary river creates hydrological, social and economic interdependencies between countries (Actionaid 2015). In this politically divided surface of the earth, more than 276 rivers and lakes are used by two or more riparian states and that pose ample scope for conflicts and cooperation over the sharing of water resources. The basins of transboundary rivers cover 46% of the earth's surface shared by 148 nations. Transboundary water accounts for 60% of the flow carried by all rivers. About 40% of the global population depend on water from their upstream countries to meet their daily use of water (Braga 2014). The uninterrupted population growth, economic development, expansion of irrigated agriculture, climate change and increasing demand for non-fossil fuel energy have contributed to ever increasing stress on water resources and ultimately threatened riverine ecosystem. The dams and reservoirs have disconnected the rivers longitudinally. The rivers are not only overexploited for irrigation and hydro-power generation but also being used to discharge polluted wastewater. Many rivers of the world are now alarmingly polluted and degraded affecting both the surrounding ecosystem and threatening the public health and livelihood of riparian population at large (WCD 2000). The impact polluted water often transmitted downstream beyond the international border. In absence of strict guideline for the rational use of transboundary water, many upper riparian states rely on the *doctrine of absolute sovereignty* in support of their claim to exploit and divert water unilaterally. The other group has faith on *doctrine of territorial integrity* which ensures fair share of water for the lower riparian states. Thirdly, the *principles of equitable utilization* were formulated to promote regional cooperation over the sharing of international watercourses. It was the beginning of the law to guide transboundary water sharing and was first formulated in 1910 by the International Law Institute. The draft was further elaborated by International Law Association and subsequently appropriated in Helsinki Rules 1966 (Subedi 2005).

11.1 Provisions in the International Rules

The issue of sharing transboundary water was addressed in Helsinki Rules (1966). The UN convention on the law of the Non-Navigational Uses of International Watercourses (1997) adopted ten general principles for the mutual and sustainable use of transboundary water. It took into account many other issues that were not paid proper attention in Helsinki Rules (1966), and also removed some confusing elements (Rudra 2017). The article V advises that States shall not only use water of an international watercourse in equitable and reasonable manner but also ensure protection of the same. It is advised that the hydro-ecology of the basin, economic need of the water, possible downstream impacts of uses of watercourses, present and future demand of water and availability of alternatives should be explored for best possible utilization of a transboundary river. It may be noted that it is obligatory for a State (vide article VII and VIII) not to degrade watercourses and also to cooperate towards promoting mutual benefits of both the human society and the ecosystem. The article IX proclaims that the data relating to hydrology, meteorology, flora, fauna and water quality are to be shared by the States occupying a basin. Since harmful effluents discharged in a river flows downstream with water, the article XXI provides guideline for abatement of pollution. In apprehension of possible impacts of pollution, the States are advised to prevent, reduce and control pollution load in a transboundary river mutually or individually so that it does not impair biodiversity and human health. The policies may be harmonized for the greater interest of both the States. The article XXIV motivates the States to ensure environmental flow adequate for uninterrupted functioning of the riverine ecosystem.

11.2 The Rivers Crossing Indo-Bangladesh Border

The partition on Bengal was one of the consequences of the independence of the 15 August, 1947, when India and Pakistan emerged as two different countries. The Bengal Boundary Commission which was constituted to delineate the boundary between West Bengal and East Pakistan (now Bangladesh) declared that '*that the Province of Bengal as, constituted under the Government of India act 1935 should cease to exist and there should be constituted in lieu thereof two new Provinces to be known respectively as East Bengal and West Bengal.*' Cyril Radcliff who was the Chairman of the Bengal Boundary Commission formed on the 30 June, 1947, found it extremely difficult to delineate a boundary which could distinctly divide the areas of Muslim and non-Muslim majorities (Bagge 1950). The commission found it quite impossible to draw a boundary without causing some interruptions in both rail and river ways. The international boundary crossed over 54 rivers creating chance for conflicts over the sharing of transboundary water resources (see Table 11.1). Many Indian rivers were tamed with the vision of irrigation and power generation during post-independence era. Any structural intervention across the river reduces downstream flow and that often leads to conflict between upper and lower riparian countries. The first conflict on the sharing of the

Table 11.1 Rivers which cross Indo-Bangladesh border

The Ganga basin	The Brahmaputra/Jamuna basin	The Meghna/Barak basin
1. Raimangal	15. Atrayee	26. Bandra/Chillakhali
2. Ichhamati–Kalindi	16. Karatoya	27. Bugi/Bhogai
3. Betna–Kodalia	17. Talma	28. Dereng/Nitai
4. Bhairab	18. Ghoramara	29. Simsang/Someswari
5. Mathabhanga	19. Deonai–Jamuneswari	30. Kynshi/Jadukata
6. Ganga	20. Buri Teesta	31. Umngi/Jalokhali–Dhamalia
7. Pagla	21. Teesta	32. Khasimara/Nawagang
8. Punarbhaba	22. Jaldhaka/Dharla	33. Umiew/Umium
9. Tulai/Tentulia	23. Torsa/Raidak/Dhudhkumar	34. Umsohryngkew/Dhala
10. Tangon	24. Brahmaputra/Jamuna	35. Umngot/Piyan
11. Kulik	25. Jinjram	36. Myntdu/Sari-Gowain
12. Nagar		37. Barak/Surma
13. Mahananda		38. Barak/Kushiyara
14. Dahuk		39. Sonai/Bardal
		40. Juri
		41. Manu
		42. Dhali
		43. Longla/Lungla
		44. Khowai
		45. Sutang
		46. Sonai
		47. Haora
		48. Sinai/Bijni
		49. Bijoya/Salda
		50. Gumti
		51. Kakri–Dakatia
		52. Selonia
		53. Muhuri
		54. Fenni/Feni

Source Rivers beyond borders (IUCN 2014)

transboundary water started with the commissioning of the Farakka barrage Project. The second in row is the conflict over the sharing of the Teesta water. Radcliffe himself foresaw that the international boundary denies holistic eco-hydrology of the Bengal. The extreme seasonal variation in flow of rivers and alarming low flow in the lean months emerged as the crux of conflict. In divided Bengal, the Ganga–Brahmaputra–Meghna system continues to flow cutting across geography, cultures and borders. The challenge for society is to cope with these hydro-geomorphic phenomena coupled with an informed knowledge base relating to the fluvial system and that may lead to a positive hydro-diplomacy (Rudra 2017).

11.3 Sharing the Ganga Water

The annual flow in the rivers of this subcontinent temporally skewed, and more than 80% of discharge passes during the period from June to September. The situation compels to store and transfer the monsoon flow to the lean season to meet the growing demands for irrigation, power generation and other needs. The structures built across the rivers impound massive volume of water, and downstream stretches are made almost dry. Thus, dams and barrages disconnect the rivers longitudinally. It has been often denied that maintaining ecological flow of rivers is critical to the sustenance of biodiversity along with the well being of millions of humans who depend on the rivers. It is urgently needed to explore a balance between the volume of water that may be extracted from the rivers and the flow to be maintained in the downstream stretches to sustain the ecosystem.

The hydro-diplomacy had always been an important issue of bilateral relationship between India and Bangladesh (erstwhile East Pakistan). The navigation in Hugli River had been extremely difficult even during the colonial period due its feeble flow and shallow depth. The British Engineers tried to divert water from the Ganga towards the port of Kolkata by excavating the off-take point of the Bhagirathi–Hugli River but failed to achieve the goal. The Government of India, after its independence in 1947, decided to build the barrage at Farakka. The issue of diversion of water from the Ganga has been debated long, initially between India and Pakistan and subsequently between India and Bangladesh. In 1962 Dr. A. T. Ippen and C. E. Wicker, the eminent hydraulic engineers who worked for the Government of Pakistan to assess the possible impacts of Farakka barrage Project opined that '*the diversion of freshwater into the Hooghly river through the construction of the Farakka barrage, will not contribute to the solution, but is likely to accentuate the serious shoaling problems in river and to the preservation of the port of Calcutta*' (Crow 1986). After the partition of Pakistan and emergence of Bangladesh in 1971, the latter raised constant voice against the Farakka barrage Project with the presumption that diversion of 40,000 cusec of water at Farakka would have many negative impacts in Bangladesh. It was anticipated that the diversion of water from Farakka barrage would jeopardize the irrigation projects, navigation in the distributaries, and allow ingress of salinity in both the distributaries and groundwater pool. The diminution of flow in rivers would affect agriculture, industry and domestic water supply. The underground water table was also expected to be lowered causing diminution of moisture content in soil.

Bangladesh demanded for equal share of the Ganga water, i.e. 40,000 cusec, and that made the situation difficult to resolve as the minimum flow in the Ganga at Farakka in the month of April was observed to be often less than 55,000 cusec. Both the countries explored for a mutually acceptable solution, and the first Indo-Bangladesh agreement was signed on the 18 April, 1975. India agreed to withdraw 11,000–16,000 cusec of water for the period of 21 April to 31 May, which was treated as the driest period of the year. Subsequently, the agreement was unilaterally nullified by the Government of Bangladesh, and subsequent agreement was signed on the 5 May, 1977. That was done based on 75% assured flow of the river at Farakka and the period taken into account was 1948–1973. The third agreement of the 7 October, 1982, was virtually a renewal of former agreement. The Table 11.2 describes Indo-Bangladesh sharing of the Ganga water as agreed in 1977. The agreement remained valid for the lean season between 1 January and 31 May and allocated volume of water for both the countries on the basis of ten days' cycle.

In 1996, India and Bangladesh embarked on a long-term agreement that would remain valid for three decades (Table 11.3). This agreement was based on the average flow in the Ganga at Farakka during 1949–1988. Unlike the treaty of 1977, the agreement of 1996 took into account long-term average flow but it was realized during execution of the treaty that real-time flow was far below the expectation. The flow of the Ganga tends to diminish with time so the long-term

Table 11.2 Indo-Bangladesh Agreement (1977) on sharing of the Ganga water

Month	Date	75% assured flow at Farakka (cusec)	Share of India (cusec)	Share of Bangladesh (cusec)
January	1–10	98,500	40,000	58,500
	11–20	89,750	38,500	51,250
	21–31	82,500	35,000	47,500
February	1–10	79,250	33,000	46,250
	11–20	74,000	31,500	42,500
	21–28/29	70,000	30,750	39,250
March	1–10	65,250	26,750	38,500
	11–20	63,500	25,500	38,000
	21–31	61,000	25,000	36,000
April	1–10	59,000	24,000	35,000
	11–20	55,500	20,750	34,750
	21–30	55,000	20,500	34,500
May	1–10	56,500	21,500	35,000
	11–20	59,250	24,000	35,250
	21–31	65,500	26,750	38,750

Source Rao (1979)

Table 11.3 Indo-Bangladesh Agreement (1996) on sharing of the Ganga water (in cusec)

Period	Average flow at Farakka (1949–1988)	Share of India	Share of Bangladesh
Jan 1–10	107,516	40,000	67,516
11–20	97,673	40,000	57,673
21–31	90,154	40,000	50,154
Feb 1–10	86,323	40,000	46,323
11–20	82,859	40,000	42,859
21–28	79,106	40,000	39,106
Mar 1–10	74,419	39,419	35,000
11–20	68,931	33,931	35,000
21–31	64,688	35,000	29,688
Apr 1–10	63,180	28,180	35,000
11–20	62,633	35,000	27,633
21–30	60,992	25,990	35,000
May 1–10	67,251	35,000	32,251
11–20	73,590	38,590	35,000
21–31	81,854	40,000	41,854

Source wrmin.nic.in/writeraddata/ind-bnd-treaty.pdf

average is bound to be a mythical figure having no compatibility with the real-time data. In fact this was anticipated by both India and Bangladesh in 1996, and accordingly both the countries agreed to share water equally if the flow at Farakka dwindles below 70,000 cusec.

The treaty of 1996 remains valid till 2026. While allocating the share of water, the growing demand of water in the future was not considered. Table 11.3 reveals that the flow at Farakka diminishes below 80,000 cusec in the late February and dwindles further at the late April,

Table 11.4 Teesta basin at a glance

State/country	Length		Basin area		Population (2011)		Population density
	In km	In %	In km^2	In %	In count	In %	In number per km^2
Sikkim	151	36.5	7039	56.9	610,577	13.4	87
Sikkim–West Bengal border	19	4.6	–	–	–	–	–
West Bengal	123	29.7	3294	26.6	1,729,899	37.9	525
Bangladesh	121	29.2	2037	16.5	2,221,550	48.7	1091
Entire basin	414	100	12,370	100.0	4,562,026	100	369

Source GoI (2007); BBS (2011)

when navigability in both India and Bangladesh faces serious challenge.

The volume of water in the Ganga again increases from the May due to the melting of snow in the Himalaya and that replenishes the river. It is to be noted that the water induced through the Ganga–Bhagirathi feeder canal during 1977–1996 had rarely been in tune with the recommended flow. The available flow went down to 16,000 cusec in the early April (Dasgupta 1996), and the navigation in and out of port of Kolkata continued to suffer. Bangladesh also suffered from shortage of water in the Ganga and related problem. Both the countries need to realize that the ensuring irrigation demand at cost of ecological services of the Ganga does not serve greater common good to the society.

11.4 Conflict Over the Teesta Water

11.4.1 The Teesta Basin

The Teesta originates from the snout of Kangse glacier in North Sikkim. It flows southwards through a deeply incised valley in Sikkim for a length of about 151 km and then delineates the border of West Bengal and Sikkim for about 19 km before it enters West Bengal and flows for 123 km in the Jalpaiguri and Koch Bihar districts. It ultimately flows 121 km in Bangladesh and finally joins Jamuna or Brahmaputra at Chandipur/Chilmari. The total catchment area is 12,370 km^2 distributed in India (83%) and Bangladesh (17%).

The Teesta River is replenished mainly by rainwater during June–September and snowmelt in April and May. The long-term average flow at Gajaldoba in Jalpaiguri district in the month of August is about thirty times larger than the flow in the month of February. Notably, demand for irrigation water goes high when the river carries lowest flow. The demand of water largely depends on the number of population living in the basin. The lower Teesta basin is densely populated. The population in Sikkim, West Bengal and Bangladesh are shown in Table 11.4.

11.4.2 Unrealistic Planning

The sharing of Teesta water has added a new dimension in the Indo-Bangladesh hydro-politics. In the 1990s, both India and Bangladesh built barrages across Teesta at Gajaldoba (India) and Duani (Bangladesh), respectively. Both the barrages were planned with unrealistic command area, and consequently project inevitably faced crisis of water shortage. The crux of the problem is that the minimum flow in the Teesta is recorded as less than 100 cumec (3533 cusec) in the first week of February but the two canals taking off from Gajaldoba barrage (India) and the one from Duani barrage (Bangladesh) were designed to withdraw 520 cumec (18,372 cusec) and 283 cumec (9998 cusec) respectively. Neither barrage is connected with reservoir and has any provision of storing monsoon water for use in the dry season. Notably, a series of hydro-power project built in Sikkim has interrupted the normal flow of the Teesta. The situation became further complicated when a domestic water supply scheme of Siliguri Municipal Corporation in West Bengal was subsequently linked with the Teesta barrage Project.

The Gajaldoba barrage in India was expected to irrigate 922,000 ha of land at final stage though the target of first substage is 342,000 ha. In 1990, Bangladesh constructed a barrage at Duani (Lalmonirhat district) with the ultimate target to irrigate 749,000 ha of land (Islam et al. 2004) In this case also, irrigation was planned based on flowing water in the river, as no reservoir was built to store and transfer monsoon water for non-monsoon season. The phase I, completed in June 1998, was expected to irrigate 111,406 ha of land. But there remained a wide gap between potential command area and the area actually irrigated in both India and Bangladesh. A barrage can help only in supplementary irrigation to *Kharif* crop during mid-monsoon breaks. The gap between expectations and achievement deepened frustration in Bangladesh adding to the stress in bilateral relationship.

11.4.3 Hydro-Politics and Negotiations

The Teesta had been an issue of Indo-Bangladesh conflict even prior to commissioning of the barrages. In July 1983, an understanding was reached on sharing of the Teesta water in the ratio of 39 (India): 36 (Bangladesh). The remaining 25% was left unallocated and decided to be shared on the basis of subsequent scientific studies.

Subsequently, Bangladesh proposed to share water in lean season (October–April) allocating 40% to India, 40% to Bangladesh and 20% to be left in the river as the ecological flow but no consensus was reached. In 2007, the Government of West Bengal agreed to release at most 25% of water from the Gajaldoba barrage but Bangladesh was not agreeable with the proposal. The Government of India opines that if the ad hoc proposal is modified taking into consideration the previously unallocated 25% in proportion to the allocated part, then India and Bangladesh may share 52 and 48% of the total flow, respectively. But the Government of West Bengal was not in the same tune with the Government of India. It is important to note that the flow in the river downstream of both the barrages is regenerated receiving apparently invisible seepage from the ground water pool. If this sharing is agreed on the basis of the total flow at the outfall of the Teesta into the Brahmaputra river, then India's share at that point would be about the same as 75% of the flow at Gajaldoba, which the Government of West Bengal claims. Thus, an agreeable formula on the basis of previous commitments may be evolved. But a mutually acceptable proposal is far away.

11.4.4 Gap in the Irrigation

The lower Teesta basin is dominated by paddy cultivation; wet variety paddy (locally called *Kharif*) is cultivated during the monsoon (June–October) and dry variety or *Boro* cultivated during the lean months (December–April). A typical *Kharif* season crop requires about 500 mm. water over its lifetime of four months and irrigation requirement of *Boro* crop is about three times higher than the *Kharif* season crop (Rudra 2009). This may be more in the areas where the porous (coarse texture) soil allows quicker infiltration of irrigation water. The Teesta basin is endowed with heavy monsoon precipitation. Except for the occasional mid-monsoon breaks, the *Kharif* cultivation does not require any irrigation through canal network. Thus, TBPs in both India and Bangladesh may not be very relevant for *Kharif* season irrigation. But in the backdrop of the vast area (922,000 ha in India and 749,000 ha in Bangladesh) targeted for irrigation under TBPs and the available water in the Teesta, both the projects appear to be unrealistic. Less than 6% of the originally envisaged area in India is actually irrigated now; the case of Bangladesh is more frustrating. It is hard to imagine how the real-time picture of demand and supply of water could have escaped the project planners.

It is revealed from simple calculations that 3 cumec of irrigation water during the field occupancy of the *Boro* crop can suffice for about 1000 ha of land (which would be about 1 cumec for most other crops). This is the requirement at the extreme end of the command area. Considering the transmission–distribution loss, the water efficiency of a dam–canal network is less than 40% observed in different parts of India (GoI 1999). Even if the efficiency of the TBP dam–canal network is 60%, irrigation to support 1000 ha of *Boro* cultivation would require induction of 5 cumec of water at the off-take of the main canal from the reservoir. Even if 80% of 100 cumec (flow in the first week of February) is diverted for irrigation, disregarding the commitment of urban water supply, only 15,000 ha can be adequately served for *Boro* irrigation. If there is no *Boro* cultivation, and the entire irrigation is for the winter or *Rabi* crop, then at most 45,000 ha can be served. Thus, adequate irrigation coverage to anything more than the currently covered 40,000 ha is an overly ambitious goal for India. It is learnt from the experience that irrigation to 922,000 ha or even 342,000 ha of land in the first phase is an impossible task. The project can only help in supplementary irrigation during wet season. The hard reality is that farmers rely on groundwater even in the command area of TBP, and consequently depletion of the groundwater table in this area in the months of January and November has been reported in the official records (SWID 2011; CGWB 2010).

11.4.5 Water for Power Generation

The fast-flowing Teesta in its upper reach renders opportunity for the hydro-power generation. Seven projects are in operation, and five more are under construction. Further, 31 hydro-power projects with small storage facilities in upper Teesta basin are in the pipeline. These run-on-the-river projects hold back a single day's water supply for a turbine and cause disconnectivity of the flow. It has been observed that if several dams impound six hours of flow during the day and release that amount during the evening hours, then there must be substantial shortage of downstream flow in the day and excess flow at night. Further, diminishing flow may jeopardize plan of flow diversion for irrigation.

The Indian TBP was planned for twin purposes of the irrigation and hydro-power generation. There is a basic conflict between use of water for irrigation and power generation. A hydro-power station requires uninterrupted flow of water, but the irrigation requires storage of water and transfer of the same from one season to other. The use of water for hydro-power generation is often treated as non-consumptive and thought that it could be subsequently used for irrigation. But the seasonal nature of the water requirement for irrigation does not ensure the continuous operation of a power plant. It was found impossible to use water released from a hydro-power station for irrigation because the reservoir-induced water passes through the barrage at late night or early morning. The limited storage capacity of Gajaldoba barrage compels to release water when pond level exceeds 114 m.

The three hydro-power plants (each with 22.5 MW capacity) are built on the Right Bank Main Canal of the Indian TBP and located on three canal falls successively. Each of the three power plants has three turbines capable of generating 7.5 MW power, and about 60 cumec flow is required at the site to run even one of the three turbines at full capacity. This quantity amounts more than half of the discharge at Gajaldoba barrage at the leanest time of the year (1–10 February), and equivalent to the winter season's irrigation requirement of about 40,000 ha. Thus, the situation makes lean season running of the power plant impossible.

11.4.6 Maintaining Ecological Flow

It is widely admitted that any structural intervention and diversion of water from a river impair downstream ecology (McCully 2001; WCD 2000). Both the Gajaldoba barrage in India and Duani barrage in Bangladesh have negative ecological impact in the downstream section.

But there is scope for reducing the adverse impact by judiciously controlling the flow. The ecological and economic services of the Teesta include preservation of biodiversity, recharge of groundwater, sediment transport, fishing, navigation, river lift irrigation, not to mention the aesthetic aspect of a natural watercourse and its surroundings.

The ecological health of the Teesta River and the surrounding ecosystem render many services towards well being of the population living in this area. The issue of ecological flow has hitherto not received due consideration in the Indo-Bangladesh negotiations except keeping 25% of water unallocated in the ad hoc agreement signed in July 1983 of the Joint River Valley Commission. But ecological flow is not a matter of arithmetic hydrology.

It is revealed from a series of studies made for the International Union for Conservation of Nature or IUCN (Dyson et al. 2003) that the flow required for non-impairment of different ecological services may vary from 65 to 95% of the natural flow. In absence of such a study for the Teesta, a cautious and conservative approach to the specification of ecological flow requirement should be taken in both the countries.

be understood that the interest of the country cannot be served by total withdrawal of water, drying up the Teesta River and allowing the groundwater reserves to deplete. The short-term benefits that some farmers may have derived since 1998 by way of canal irrigation are not sustainable. The demand management could provide better solutions in the long term than supply side management. The sharing water of the Teesta should not be treated as an issue of arithmetic hydrology rather it is a matter of holistic ecohydrology. The downstream area now faces the peril of enormous sedimentation and continued groundwater depletion. This menace can be checked by reversal of the policy of indiscriminate diversion of lean season flow of the river. There can be no compensation for a continuing catastrophe faced by the afflicted community. It is important to explore a compromising point between the volume of water that may be diverted from the Teesta and the flow to be allowed in the river to sustain the ecosystem services. This seems to be the only way to achieve mutually acceptable solution of the Indo-Bangladesh conflict over the sharing of the transboundary waters.

11.5 Two Barrages: Myth and Reality

The TBP both in India and Bangladesh were primarily planned for irrigation during the *Kharif* or monsoon season, but subsequent attempts were made to ensure irrigation to *Boro* or dry variety paddy. But the flow of the Teesta diminishes to such a level during the summer that only 5% of the originally planned area has been brought under irrigation in the Indian side. The experience of Bangladesh has been equally bad. These setbacks can be attributed mostly to faulty planning, and not so much to under-performance. The targets set for the project were never achievable, and performance may have been in line with what had been achievable. It needs to

References

Actionaid (2015) Blues beyond boundaries; transboundary water commons/ India report. Natural resource knowledge activist hub—A knowledge Initiative of Actionaid India. Bhubaneswar

Bagge A (1950) Report of the international arbitral awards. Boundary disputes between India and Pakistan relating to the interpretation of report of Bengal Boundary Commission. Part I Published by UN (available online)

BBS (2011) Bangladesh data sheet, Bangladesh Bureau of Statistics. http://www.bbs.gov.bd/WebTestApplication/userfiles/Image/SubjectMatterDataIndex/datasheet.xls

Braga B (2014) Water without borders: sharing the flows? In: Pangare G (ed) Hydrodiplomacy. IUCN, New Delhi

CGWB (2010) Ground water scenario of India 2009–10; Central Ground Water Board Ministry of Water Resources Government of India Faridabad 2010. http://www.cgwb.gov.in/documents/Ground%20Water%20Year%20Book%202009-10.pdf

Crow B (1986) Sharing the Ganga. In: Farakka-A Gordian Knot, Ishika, Kolkata, pp 170–171

Dasgupta A (1996) Bharat-Bangladesh Jalbantan Chukti (1996): Ek Eitihasik Sambhabana, (in Bengali) Ganashakti, 16th December, 1996: 4

Dyson M, Bergkemp G, Scanlon J (eds) (2003): The essentials of environmental flows. Water and nature initiative, international union for conservation of nature and natural resources, http://moderncms.ecosystemmarketplace.com/repository/moderncms_documents/iucn_the-essentials-of-environmental-flows.pdf. Accessed 30-12-11

GoI (1999) Report of the national commission for integrated water resources development plan, ministry of water resource new. Govt. of India, Delhi

GoI (2007) Review of performance of hydropower stations 2006–07, Ministry of Power, Central Electricity Authority, Hydropower Planning and Investigation Division

Islam N, Azam A, Islam QR (2004) Teesta river water sharing: a case study in Teesta Barrage Project. In: Presented at 2nd Asian regional conference of the international commission on irrigation and drainage, 14–17 March, Moama, NSW, Australia; also available online. http://watertech.cn/english/islam.pdf

IUCN (2014) Rivers beyond border. India Bangladesh Transboundary River Atlas

McCully P (2001) Silenced river: ecology and politics of large dams. Zed Books

Rao KL (1979) India's water wealth. Orient Longman, Kolkata

Rudra K (2009) Water resource and its quality in West Bengal, WBPCB

Rudra K (2017) Sharing water across Indo-Bangladesh border. In: Bandyopadhyay S, Torre A, Casaca P, Dentinho T (eds) Regional cooperation in South Asia, Springer, Berlin, pp 189–207

Subedi SP (2005) International watercourses law for 21st century. The case of river Ganges Basin. Ashgate Publishing Limited, UK

SWID (2011) Unpublished records, state water investigation directorate, Government of West Bengal

WCD (2000) Dams and development: a new framework for decision-making, Report of the world commission on large dams. http://www.dams.org/

The Concept of Ecological Flow

12

Abstract

The importance of flowing water in the ecosystem was realized when the issues of dying rivers and the imminent threat to freshwater supply came to fore. Subsequently, there were efforts to understand the limit of maximum permissible withdrawal of water from a river so that its hydro-ecological functions are not drastically hampered. The term ecological flow means not only the quantity but also the quality of flowing water and also its cultural and aesthetic value. It identifies the required quantity, quality and distribution of flow patterns from the source to mouth of a river, preserving the life in and around the channels. The drastic reduction in flow in most of the river has the ecological cost. The already identifiable indicators of the ecological damage are longitudinal disconnectivity of the flowing channel, the falling level of the groundwater table and the loss of biodiversity. The greater common good of the human society does not lie in the abuse of the rivers; the future of civilization depends on fixing critical limits for the exploitation of water sources and allowing the rivers to flow.

The twentieth century witnessed an exponential growth of economic activities along with the worldwide abstraction of water, causing enormous damages to the ecological services of flowing water. The use water resource was treated as exclusive right of the human in the traditional engineering framework. Any water flowing into the oceans and lakes, or into a foreign territory, was regarded as wastage. In those days, it was not understood that a flow renders enormous, but often invisible, ecological services. When irreversible damages to the environment caused by river projects were noticed, the need to preserve a certain quantity of water in the river was realized. The water needed to sustain the diversity of aquatic life and the functioning of the ecosystem of a river or stream is broadly referred to as the 'ecological flow' of that river. The concept of ecological flow is still in a theoretical stage and going through validation and refinement as well as inputs from the inter-disciplinary knowledge base (Bandyopadhyay 2011).

Geomorphologically, the ecological flow may be looked upon as the volume that can maintain a horizontal and longitudinal connectivity of water in the channel even during the lean season. Horizontal connectivity means the linkage between the river and its floodplain. While dams and barrages intercepted on the longitudinal connectivity, the flood control embankments were responsible for delinking the river from its floodplain. Longitudinal connectivity refers to a level of flow that ensures nominal navigability along the length of the river. However, such concepts have only a limited use. In fact, many tropical rivers do not have any horizontal connectivity during the lean months of a year. The scientists working on the ecological flow were faced with the reality that the flows of most of the world's major rivers were already disrupted and that there are little contemporary data on 'natural' or 'unrestricted flow'! This is amply demonstrated by the fact that in this age of information, a search on the Internet produces hardly any data set on the directly measured natural flow of a major river. In the absence of a *natural* flow, they turned to model-based predictions of a *naturalized* flow that would have occurred in the absence of dams and barrages, water supply diversions and other types of water management activities that are reflected in the present-day flow.

On the other hand, it was found that the restoration of health of an already damaged ecosystem could be undertaken with the vision of *ecological integrity* which allows hydrological functioning of the river and supports survival of flora and fauna in the altered situation. From this consideration, the ecological flow was viewed as a flow regime that would ensure the protection of the riverine ecosystem. The restoration of a system to a possibly natural setting, rather than the preservation of a natural environment, became the ecologist's goal in an impaired world. New research tools were sought. New phrases such as *ecodeficit* (i.e. the reduction of flow in-stream as a consequence of large-scale extraction of water in a regulated channel.) were introduced (Homa et al. 2005). However, finding ways of ensuring the ecological integrity of a river-based ecosystem or making up the *ecodeficit* has proved to be more difficult than a coining of terms.

Subsequently, there were efforts to understand the limit of permissible withdrawal of water without impairing hydro-ecological functioning of the river. It became clear that the integrity, health and ecological services of the flowing water cannot be maintained without this understanding. The concept of ecological flow evolved in the process of these studies. It identifies the required quantity, quality and distribution of flow from the source to mouth of a river, preserving the life in and around the channels. It also recognizes the importance of environmental flows 'as a vital contributor to the continuing provision of environmental goods and services upon which peoples' lives and livelihood depend' (WWF 2012).

Since time immemorial waters in rivers and springs were said to be pure or holy, which provided the ground for the dawn of civilizations worldwide and are still the major suppliers of water for domestic, agricultural and industrial consumption. Although rivers contain only a very small percentage of the total water of the earth, they are virtually terrestrial part of the hydrological cycle and continuously transfer water and nutrients from the land to the sea or lake. The rivers together drain two-thirds of the surface of the earth. River water is used in various sectors of human life. It provides drinking water, ensures irrigation and sustains settlement along its banks. Water provides the cheapest mode of transport and is also the base for aquaculture and fishing. River flows are also the major source of freshwater for industries and carry off the wastes. With every passing flood, a river deposits fertile silt and sand. Rivers provide nourishment for numerous aquatic lives. Thus, channelled surface water flows are pivotal to many apparently invisible issues important for the sustenance of the society, and interdisciplinary researches are being conducted. The preservation of sustainable flow in rivers and streams is of utmost importance to protect the ecosystem services.

The water sustains riverine ecosystems, often overflowing, submerging the adjacent floodplains and in turn exchanging dissolved and particulate matter with the surrounding land. Life within a

12.1 River Ecology

The importance of water in the environment was underscored when the issues of depleting water resource and the imminent threat to freshwater supply came to fore (Gleick Peter 1993).

river system largely depends on the local pattern and volume of flow, geology, geomorphology of the stretch, temperature, salinity, pH factor, toxicity, dissolved and particulate nutrients and sediments, etc. Microbial, floral and faunal character of a river exhibits progressive differences from tributary to distributary channels. Therefore, any study on the ecological or flow character of the river must be scale specific. The river flow, energy, biotic interactions and water quality may be also dependent upon the temporally variable flow, recurrence interval of a particular flow regime, its magnitude, timing and how fast changes from one hydraulic condition to another. A change in one of the factors triggers interrelated changes in the ecosystem (Closs et al. 2004).

Temperature greatly affects aquatic ecosystem. Dissolved oxygen in the water is affected by temperature, turbulence and organic content. As temperature rises, the gas-absorbing capacity of water decreases rapidly, reaching zero at 100 °C. It should be added that availability of oxygen is also a limiting factor influencing the distribution of the species. Fast-flowing water due to higher turbulence tends to be saturated with oxygen compared to stagnant water which contains less oxygen, especially in the winter season. It is either produced by photosynthetic exchange from aquatic plants or by diffusion at the surface between air and water. A turbulent flow often enhances the solution of dissolved gases such as oxygen, and concentrations may even reach 100% (Closs et al. 2004). The churning action makes the water saturated with air and brings about homogeny of pH, temperature and dissolved substances.

The flow also impacts the riverine ecology, particularly near the surface. Its erosion washes away lighter soil particles, finely divided organic matter or detritus and small floating living organisms. Planktons do not favour the rapidly flowing streams and tend to accumulate on obstructions. Nutrient spiralling causes minerals and organic matter to be delayed in being lost or transported downstream. This, in turn, allows the nutrients to cycle within the food chain of aquatic organisms. Species accumulation can be highly variable and unstable owing to the swift currents and force of water drifting away the organisms. The deltaic rivers tend to be turbid, due to the accumulation of finer suspended load. Tropical rivers tend to have greater turbidity due to intense torrential rainfall, compounded by human intervention in the form of deforestation, topsoil loss due to agriculture and so on. In the lower reach of rivers, tidal action increases turbidity even further. Turbidity impairs sunlight penetration restricting aquatic plants and phytoplanktons from photosynthesizing. The suspended sediments not only block the respiratory surface of the fish gills but also act as a carrier of various chemicals harmful to the aquatic fauna. Some aquatic species thriving in the non-tidal stretch of a river may not survive in such conditions.

The flow of a river may be intermittent or perennial. Even perennial rivers can have substantial temporal variation in the flow in a year. This induces seasonality in the ecological conditions of its surroundings. The diversity of species community is maximum at the river bed. In stream ecology, the channel-bed acts as the site for aquatic organisms to lay and incubate eggs and also offer shelter during floods (Minshall 1984; Statzer et al. 1988). It forms a refuge for benthic organisms in time of floods, droughts, etc. Ward and Stanford (1983) called it a 'faunal reservoir', capable to cope if flora and fauna are threatened by hostile situation. The stream-bed being rich with organic matter provides nutrients for primary consumers. The biologically active region between the bottom of river beds and groundwater region is known as the *hyporheic* zone. Here, organisms create their distinct habitat within the interstices (Gordon et al. 2004). It has been also said that higher the coarseness of the gravels at the channel bed more will be the micro-organic activity and biodiversity. Thus, a sharp rise in fine sediment load could fill up the inter-gravel spaces clogging the invertebrate habitats, trapping the eggs, and hence, a drop in the biodiversity may be observed.

The movement of freshwater organisms may happen either by swimming through the water, treading along the river bed, floating on the water surface or by drifts. Thus, water currents, wind and waves determine the mobility of organisms. The dispersal of organisms, small plants, seeds, spores, stems, leaves capable of vegetative reproduction and so on helps in the species regeneration. *Passive* dispersal is caused when organisms or plant parts are dispersed entirely by water currents and settle on the river bed/bank once the flow subsides. *Active* dispersal is caused when invertebrates control their movements and make choices of flow distance, direction and depth by swimming.

Estuaries are regions where tidal interaction causes the saline sea water to mix with the freshwater from the upstream, resulting in fluctuations of temperature, salinity and turbidity. The Sundarban is a classic example of the estuarine ecosystem. The recent reduction of the freshwater supply and resultant increase in the salinity have impaired the Sundarban ecosystem. The discharge of the polluted water into the Kultigong at Ghusighata has further aggravated the situation. The estuarine ecology has attracted many studies worldwide. It is dominated by fine sedimentary materials forming mudflats. It is one of the vital breeding sites for many species of birds; marine and anadromous fishes (salmon, *hilsa*) and encourages the luxuriant growth of unique mangrove ecosystem. The rising levels of salinity, increasing sea-temperature and muddy substrate, cause 'physiological stress' in organisms such as clogging of gills and malfunctioning of organs and a tendency to change their habitat. The dumping waste into the estuarine system or overexploitation of the available resources causes changes in nutrients, productivity, biodiversity, biomass distribution, bioaccumulation of contaminants and instability of the overall biotic and abiotic compositions (McLusky and Elliot 2004).

12.2 Impact of the Altered Flow Regime

The flow of a river is generally altered by the construction of dams and barrages. Such structural intervention changes the physical or chemical characteristics of water and brings about a marked change in the biodiversity of the river and only those organisms, which are able to adjust to the changing environment survive. The longitudinal changes in species-community structure may vary with changes in food availability and the predator–competitor relationship within the ecosystem. Viewed from within that structure, human intervention on the flow regime generally comes in the shape of a sudden disruption of the normal functioning of the system. The damages are often so irreversible that even after the withdrawal of that intervention, the biotic community may not be able to re-establish itself. For example, the Damodar River is made so dry downstream of the Durgapur barrage that many species have disappeared forever.

The biotic life in upstream and downstream of a dam or a barrage differs markedly. Since the commissioning of the Farakka barrage, the anadromous *hilsa* could not penetrate the fish ladder designed through the barrage and reach further upstream. Grievances of fishermen in Bhagalpur for the lowering in catch are evidence to this fact. Substantial change in the fluvial regime of a river may also occur in the absence of a control structure. Tropical rivers recharge the groundwater pool during the rainy season and are fed from that pool during the lean months. Thus, the flow in the river is only the visible part of a system of flows distributed over a much larger region. Intense use of groundwater may disrupt this flow, reducing in turn the lean season flow in the river. Another form of intervention that does not require a structure is loss of topsoil due to deforestation and/or agriculture with heavy tilling.

Apart from affecting the quantity of flow, human intervention can also drastically alter water quality, through mixing of pollutants which degrade the physic-chemical properties of the flowing water and ultimately threatens the biodiversity (Collocott and Dobson 1974). Rivers are frequently used as conduits of wastewater disposal. As long as the population density is low and the area is non-industrialized, the ecosystem can generally cope with the volume of wastewater generated, through a process of self-recovery. The sources of pollution may be classified as non-point, such as agricultural run-off and point, such as through pipes and drains carrying sewage. The stretch of the Hugli River in West Bengal between Kalyani and Diamond Harbour has been polluted by more than three hundred point sources carrying both domestic and industrial wastewater.

Many rivers of Bengal were embanked to control flood since the mid-nineteenth century. While floods pose destructive hazards, they are also naturally occurring extreme events that influence the ecological processes of the channel and its surrounding environment. It is an essential part of the overall health of the river and its floodplain. A flood can play a crucial role in preserving the longitudinal connectivity of the river, by flushing out sediment deposition. The flood helps the geomorphic and ecological functioning of the stream including sediment and nutrient movement and maintaining habitat conditions. It has many ecological roles such as removing toxic elements and thereby reducing the biological oxygen demand (BOD) levels and increasing the dissolved oxygen (DO) levels. The flood deposits fertile silt over the floodplain, thus restoring the hydrological regime and fertility of the soil. It recharges the groundwater table and ensures the base flow into the river during lean seasons. The flood also facilitates flourishing of the aquatic community in the water-filled pockets on the floodplain. Some aspects of the relation between river flow and the surrounding ecosystem are only beginning to be understood. We should stick to the doctrine of minimum intervention till the delicately balanced hydro-ecology of the GBM delta is fully understood.

12.3 Defining the Concept of Ecological Flow

An assessment of ecological flow must begin with a research on the temporally variable natural flow in the various parts of a river. Historic hydrological data on unrestricted river are the basic ingredients of such a study. Such time series data reveal not only the average flow pattern in different sections of the river, but also the range of 'normal' variations that the river ecosystem has experienced historically. In addition, these data help one to understand the extent of flow corresponding to naturally occurring extreme events such as drought and flood, which have a role in shaping the river landscape and ecosystem. Indeed, there have been attempts to specify the ecological flow requirement as a percentage of the unrestricted river flow.

Other approaches of quantifying ecological flow have centred on different aspects of a river system. There are hydraulic methods that relate to river cross section, wetted perimeter and hydraulic geometry calculations. Habitat methods measure the benefits of in-stream flow to aquatic habitat and relate hydraulic properties of a river to the biological requirements of certain species (Homa et al. 2005). The in-stream flow assessment method attempts to combine the historic stream flow methods, hydraulic methods and habitat methods (Jowett 1997).

The restoration of health of an already damaged ecosystem could be undertaken by using the notion of *ecological integrity*, which includes capability of a river to allow organisms to function in a way *comparable to* that of the unaltered state (Karr and Dudley 1981). From this consideration, ecological flow was viewed as a flow regime that would protect ecological integrity of a river system (NC 2009). Restoration of a system to a possibly natural setting, rather than preservation of a natural environment, became the ecologist's goal in an impaired world.

12.4 An Engineering-Management Approach

Policy-makers were also confronted with the challenge of rectifying a domain that has become non-natural, without precisely knowing what the natural scenario could have been in the absence of human intervention. As a result, their thinking has been rooted to what is precisely known and perfectly visible, namely the intervention structures and the current status of water use. It is now understood that the minimum flow required for hydro-ecological functioning of the river needs to be maintained. Such a policy is easier to implement in the absence of an intervention structure, though such structures are rather common in this subcontinent. Therefore, many experts suggest that flows that take place as a result of the seepage from groundwater table in an unaltered condition can be maintained by routing flow from the reservoir. The artificially bolstered flow regime aims at restoration of the natural flow in the channel and protection of the aquatic ecosystem. It considers flow magnitude, duration, frequency, timing and variability in individual stretches of the channels, as well as the requirement of a multitude of species within the ecosystem. In other words, the goal is to quantify natural conditions in terms of a number of characteristics and to simulate those conditions as much as possible in a controlled environment. By and large, policy objectives of Governments around the world have been set along these lines. For them, the question of ensuring ecological flow has specialized into finding appropriate ways of operating engineered structures with control mechanisms, with some consideration given to ecological aspects.

This reduction of the problem has given rise to proliferation of solutions from the field of management. In the past, water resource management strategies sought to strike a balance between competing demands of water for irrigation, hydro-power generation, water supply to cities and industries and so on. In view of the recent prominence of ecological concerns, the list of considered factors has been expanded to include a number of ecological factors. The term *integrated water resources management* has been coined to reflect this expansion/diversification. In fact, the modern managers have been rather liberal in expanding the list. The new list includes recreation, cultural and spiritual needs of people, which ecologists had never thought about. The intended balance may be attained by reducing the abstraction, regulating release of water from the reservoir and planning demand-side management (Land and Water Australia 2007). The term *environmental flow* refers to those parts of dams-induced water which is dedicated exclusively for environmental benefit in contrary than what is released for irrigation, hydro-power generation or domestic consumption. It is also used in a slightly broader sense, so as to include non-withdrawal or deferred withdrawal with environmental objectives. The environmental flow objectives include all the recent additions to the conventional objectives, namely,

 i. To conserve the river regime,
 ii. Allowing the self-rejuvenating mechanism of the river to function,
 iii. Conserving in-stream biodiversity,
 iv. Recharging groundwater,
 v. Facilitating livelihoods,
 vi. Maintaining sediment movement,
 vii. Preventing saline intrusion in estuarine and delta areas,
 viii. Allowing the river to fulfil the cultural and spiritual needs of people,
 ix. Providing recreation.

12.5 A Rational Meeting Point

The engineering-management approach described above is based on the presumption that some abstraction/diversion of water is unavoidable for meeting human needs, even though no abstraction is best for the environment. It is envisaged that the flow would be regulated up to a level agreed upon by the stakeholders, so that the ecosystem may function in a degraded manner, but would not face the threat of total extinction (Bandyopadhyay 2011). A negotiated level of degradation may not be sustainable. The stakeholders' perceptions are based on aspirations and anxieties of the present day, which can come in the way of a candid view of the matter. The Hoover Dam was built on the Colorado River at a time when there was a measure of shared intoxication about the wonders of development projects. The massive level of intervention there did not generate any discontent among stakeholders. Yet the project ensured the premature death of the Colorado River, which failed to reach the Gulf of California. The consequence was one of the greatest ecological disasters of the twentieth century.

12.5 A Rational Meeting Point

The risk of non-sustainability of the engineering-management approach stems from its perspective, which can arguably be referred to as a dam-side view. This perspective treats the diversion/abstraction interests as indispensable and in conflict with the ecology and looks for a peaceful resolution of that conflict through compromise. However, the appearance of conflict may have resulted from a limitation of the perspective itself. Ecology encompasses us all. Humans are only a part of ecology. Our well-being is contingent on an intricate combination of finely balanced conditions, which we do not fully understand. Natural or other variation in those conditions offers us an opportunity to study the consequences of these variations. Often long-term effects are noticed after much damage has been done. For example, the chilling fact of all-round depletion of the groundwater pool in the Gangetic plains was not foreseen during the heady days of green revolution, and by the time its magnitude and significance were understood, the momentum of that depletion had become too large to control. The history of these setbacks may temper our confidence in the technological as well as management capabilities. Any model-based decision involves simplifying assumptions. As shortcomings of models become apparent, better models and better decisions do emerge. However, correcting past mistakes does not reduce our natural propensity to make new mistakes, as our understanding is never complete. Thus, human intervention of any sort cannot be allowed to exceed the range of normal variations observed over a protracted period. Until such time as when the long-term impacts of small interventions are well understood, one has to tread on the side of caution. This approach has apparently been adopted in South Africa (Silk et al. 2000).

In an IUCN compilation of such studies (Dyson et al. 2003), it has been noted that the range of flow required for non-impairment of different ecological factors varies widely. Some indicators can withstand 65% of natural flow, while others require 95% of the total flow. Studies in Queensland, Australia, suggest that around 80–92% of mean annual flow may be required to avoid a low risk of environmental degradation (Arthington and Pusey 2003). An Expert Reference Panel advising on environmental flows for the River Murray used benchmarking to recommend 65–75% of natural flow that can save the ecosystem from unacceptable environmental degradation (Jones et al. 2002). Greater levels of flow disturbance resulted in a range of different river conditions.

The outcomes of studies taken up so far in different parts of the world underscore the fact that each ecosystem is different. As far as the understanding of a specific river system is concerned, there is no alternative to intense and location-specific study on different aspects of the river and its ecosystem. While the results of such a study are awaited, a cautious and conservative approach to specifying ecological flow requirement should be taken. From this consideration, and based on the IUCN experience (Dyson et al. 2003), it would be prudent to permit at least 95% of normal average flow in the channel, and it would be a dangerous gamble to permit anything less than 80% of the normal average flow.

The exercise of caution on the supply side of water resources would inevitably lead to questions of demand management. This is difficult to achieve, as individual users are often broadly dispersed, and there is a lack of scientific understanding of the functioning of an ecosystem. Yet, this understanding must reach the spots from where water demand is generated. An appreciation of the natural world around us cannot be the exclusive privilege of the experts. The disconnection between humans and ecology lies at the root of accumulation of human demands to a scale that threatens to destroy the ecology. The tempering of that demand is crucial to the feasibility of ensuring the ecological flow through a river.

12.5.1 The Indian Scenario

Ecological considerations did not enter the river-planning scenario of India until the late twentieth century. The National Commission for Integrated Resource Development (NCIWRD

1999) estimated the overall water requirement for 'environment and ecology' to be about 1% of the total national water requirement. No geographical or seasonal distribution of this bulk requirement was given nor was there any rational justification. The matter of a minimum level of flow in rivers gained prominence in the 14 May 1999 order of the Apex Court of India in the Sureshwar D. Sinha vs. Union of India case [W.P. (C) 537 of 1992], which said, on the basis of findings of an expert committee, 'a minimum flow of 10 cumec (353 cusec) must be allowed to flow throughout the river Yamuna' (Dutta 2009). In 2001, the Government of India decided to constitute the Water Quality Assessment Authority (WQAA) to explore the critical flow in rivers to protect the ecosystem. The expert committee considered many issues including the temporal variability of flow, critical trade-off between different sectoral demands and environment, cost of mechanical treatment of the wastewater, sharing of transboundary water and suggested for assessing a 'minimum flow' which can maintain desired water quality. Iyer (2005) noted that the use of the term 'minimum flow' indicates the intention of maximum abstraction. It reflects a mindset that somehow regards a natural resource like water as a state-owned asset, to be partially released as a central assistance or subsidy extended to the environment—in the spirit of minimum support price for rice and wheat. Amarasinghe et al. (2005) attempted to estimate the environmental water demand of India, aggregated over nineteen major river basins, by using simulated data from a global hydrological model (Smakhtin et al. 2004). The authors arrived at the figure of 476 cubic kilometres (about 42% of the annually utilizable water resources in India) for the estimated annual volume of ecological flow. The global hydrological model was not calibrated with any hydrological or ecological data from India. Smakhtin and Anputhas (2006) used two methods, based on shifting of *flow duration curves* or FDC (Petts 1996) and *desktop reserve model* or DRM (Hughes and Münster 2000; Hughes and Hannart 2003), respectively, to classify rivers into a few *environment management classes* (EMC), and used them to asses environmental flow requirement in some major Indian rivers, on the basis of limited hydrological data. Their findings are summarized in Tables 12.1 and 12.2.

The tables attempt to provide the minimum flow (as percentage of mean annual run-off) that would leave the river in a particular environmental management class. The desired management class is supposed to be taken through informed negotiation among stakeholders and that may not be the same as the present status of the river. The small percentage of environmental flow requirement for the 'natural' class (class A) in the cases of many rivers, according to Tables 12.1 and 12.2, is notable. The authors admit that their methods need more refinement with inputs hydrologists, ecologists, water managers and other scientists and say that their findings should not be regarded as prescriptions for environmental flow in India or elsewhere. Smakhtin et al. (2007) provided a framework for studying present hydro-ecological status of Indian rivers basins, on the basis of Table 12.1, and used it to specify the *present* environmental management class of certain sections of the Tungabhadra, Krishna, Kaveri and the Ganga basins, using some currently available ecological data for those basins. None were found to be in class A (natural).

In February 2009, the Government of India set up the National Ganga River basin Authority (NGRBA) 'to ensure effective abatement of pollution and conservation of the river Ganga for a comprehensive planning and management'. In June 2010, NGRBA entered into a Memorandum of Understanding with an expert group comprising seven IITs, asking it to produce a Ganga River basin Management Plan that would include specification of a minimum flow in the river for ecological/environmental purpose. The expert group submitted its final report in 2015 and defined environmental flow as a *'regime of flow in a river or stream that describes the temporal and spatial variations in quantity and quality of water required for freshwater as well as estuarine ecosystems to perform their natural*

Table 12.1 Estimated environmental water requirement (EWR) (in per cent of *mean annual run-off* or MAR) at outfalls of major Indian rivers for different environmental management classes obtained using FDC shifting method

River	Natural MAR (BCM)	Present-day MAR (BCM (% of MAR))	EWR (% natural MAR)			
			Class A (natural)	Class B (slightly modified)	Class C (moderately modified)	Class D (largely modified)
Brahmaputra	585		78.2	60.2	45.7	34.7
Kaveri	21.4	7.75 (*36.2*)	61.5	35.7	19.6	10.6
Ganga	525		67.6	44.2	28.9	20.0
Godavari	110	105 (*95.4*)	58.8	32.2	16.1	7.4
Krishna	78.1	21.5 (*27.5*)	62.5	35.7	18.3	8.4
Mahanadi	66.9		61.3	34.8	18.5	9.7
Mahi	11.0		41.9	17.1	6.5	2.3
Narmada	45.6	38.6 (*84.6*)	55.5	28.8	14.0	7.1
Pennar	6.3		52.7	27.9	14.3	7.3
Tapi	14.9	6.5 (*43.6*)	53.2	29.9	16.6	9.0
Periyar	5.1		62.9	37.3	21.2	12.1
Sabarmati	3.8		49.6	24.2	12.1	6.6
Subarnarekha	12.4		55.0	29.9	15.4	7.4

Table 12.2 Estimates of long-term environmental water requirement (EWR) volumes (expressed as % of *mean annual run-off* or MAR) for selected river basins and different environmental management classes obtained using DRM method

River	Site	Mean annual run-off (BCM)	Hydrological variability index	Long-term EWR (% MAR)			
				Class A (natural)	Class B (slightly modified)	Class C (moderately modified)	Class D (largely modified)
Brahmaputra	Pandu	573.8	1	85.4	54.5	40.6	38.6
Kaveri	Krishnaraj Sagar	5.37	3.4	50.8	35.8	26.7	21.7
Ganga	Farakka	380	1	82.4	52.9	39.7	38.1
Godavari	Davlaishwaram	96.6	4.7	45.4	32.7	24.5	19.6
Krishna	Vijayawada	56.7	5.8	38.4	27.8	20.8	16.1
Mahanadi	Hirakud/Sambalpur	54.8	5.1	44.7	32	24.1	19.2
Mahi	Sevalia	12.2	13.7	32.7	23.3	16.9	12.3
Narmada	Garudeshwar	22.6	5.4	37.3	26.8	20	15.5
Pennar	Nellore	2.34	7.7	33.3	24.7	18.7	14.3
Tapi	Kathore/Ghal	4.5	6.7	36.9	27.6	21.4	17
Periyar	Planchotte	5.15	4.6	38.1	27.1	20.1	15.7
Sabarmati	Ahmedabad	1.04	8.6	34.1	24.7	18.2	13.6
Subarnarekha	Kokpara	9.76	8.1	35.3	25.6	19	14.6

ecological functions (including sediment transport) and support the spiritual, cultural and livelihood activities that depend on these ecosystems' (Tare et al. 2015). This definition is commensurate with the engineering-management framework. The natural flow of the Ganga River occupies a central place in the spiritual and cultural ethos of northern India, and the qualitative and quantitative degradation of flowing water has often given rise to spontaneous discontent among the public. The approach adopted in the above definition, as well as in some preceding work (Sinha and Prasad 2005), seeks to internalize this sentiment by reducing spiritual and cultural aspects into measurable quantities such as the water depth needed for *sadhu*s to bathe or for non-depreciation of religious places. A similarly simplistic approach is also adopted in the report of a project for the assessment of environmental flows of the upper Ganga basin, undertaken by the World Wildlife Fund (WWF 2012) under its Living Ganga Programme. In this report, the current levels of flow at various sites along the river were pegged at 18–25% smaller than the 'natural flow' simulated from a hydrological model. Recommendations were generally in favour of large-scale abstraction in the monsoon months and some augmentation of the lean season flow.

The thinking about ecological flow in the Indian scenario has been almost entirely dominated by the engineering-management approach and much work remains to be done. The actual flows in the rivers need to be measured. The Central Water Commission has gauge stations at numerous locations for measuring flows of small and large rivers all over the country. That data are inexplicably kept confidential. As the data never receives the normal scientific scrutiny, it often remains faulty. Unfortunately, the entire data collected by CWC, at considerable public expense, remain inaccessible to the public and unusable by the scientific community. This national wastage has to be checked, and the data must be released, and the data collection mechanism has to be exposed to scientific scrutiny. The run-off regimes in large basins do not give a good idea about the status of its various parts. The sub-basin level run-offs need to be tracked and modelled, and ecological studies also need to be undertaken.

12.5.2 Experience of Bengal

It is widely noted that the rivers of Bengal are traditionally double channelled. The monsoon peak flow establishes lateral connectivity from bank to bank and even spills over on the adjoining floodplains, extending far beyond, delimiting the actual width required by the channel during flood. This is identified as the space for the river. Depending on the volume of flow during the monsoon, wavelength and amplitude of the meandering change. The river tends to alter its wavelengths and amplitudes when a larger volume of water flows through it and shrinks to a feeble channel with dwindling discharge. Thus, a tropical river has a narrow channel within a wider channel. The larger one is to accommodate the monsoon flow, and other is for the lean season.

The ever-increasing dependence on groundwater for irrigation and domestic use has impacted the flow in rivers. Excessive extraction has already led to alarmingly low levels of groundwater. The base flow which earlier replenished the lean season flows in river has substantially reduced. Most of the rivers in Bengal go dry during the lean season, disrupting both longitudinal and horizontal connectivity. Furthermore, catchment deforestation, paving of the surface, mining and other industrial activities have altered run-off-infiltration ratio and especially in the western plateau region of West Bengal, where the lateritic tracts do not allow easy percolation of water into the subsoil. The priority of irrigation over preservation of the ecological flow has caused the river Damodar to go dry downstream of Durgapur barrage. Same is the case in Mayurakshi, downstream of Tilpara barrage or that of Kasai downstream of Mukutmanipur reservoir. The Teesta barrages in both India and Bangladesh operate on same logic of exploiting maximum possible water for irrigation and thus make the downstream stretches dry (Figs. 12.1 and 12.2).

There has been no quantification of ecological flow in the rivers Bengal. The lean season's flows in rivers of North Bengal have substantial

Fig. 12.1 Dry bed of the Damodar downstream of Durgapur barrage

Fig. 12.2 Left bank canal that withdraws water for irrigation

snow-melt component, and the flow in the rivers of south Bengal is ensured by groundwater pool. The drastic reduction in flow in most of the rivers has an ecological cost. A striking indicator of the ecological damage is the desiccation of many rivers in South Bengal. The ecological flow requirements in these rivers need to be investigated. Better estimates of natural flow, possibly based on observed run-off data, would facilitate quantification of the ecological flow.

References

Amarasinghe UA, Shah T, Turral H, Anand BK (2005) India's water future to 2025–2050: business-as-usual scenario and deviations. Research Report 123, International Water Management Institute

Arthington AH, Pusey BJ (2003) Flow restoration and protection in Australian rivers. River Res Appl 19(5–6): 377–395

Bandyopadhyay J (2011) Deciphering environmental flows. Seminar (October Issue) 626:50–53

Closs G, Downes B, Boulton A (2004) Freshwater ecology. Blackwell Sc Ltd.

Collocott Thomas Charles, Dobson Alan Binaloss (1974) Chambers dictionary of science and technology: L-Z, vol 2. W. and R, Chambers

Dutta Ritwick (2009) The unquiet river—an overview of select decisions of the courts on the River Yamuna. PEACE Institute Charitable Trust, Delhi

Dyson M, Bergkemp G, Scanlon J (eds) (2003) The essentials of environmental flows. In: Water and nature initiative, international union for conservation of nature and natural resources. http://moderncms.ecosystemmarketplace.com/repository/moderncms_documents/iucn_the-essentials-of-environmental-flows.pdf. Accessed 30 Dec 2011

Gleick Peter H (1993) Water in crisis: a guide to world's fresh water resources. OUP, Oxford

Gordon ND, McMahon TA, Finlayson BL, Gippel CJ, Nathan RJ (2004) Stream hydrology: an introduction for ecologists, 2nd edn. Wiley, Chichester

Homa ES, Vogel RM, Smith MP, Apse CD, Huber-Lee A, Sieber J (2005) An optimization approach for balancing human and ecological flow needs. In: Proceedings of the EWRI. World Water and Environmental Resources Congress, ASCE, Anchorage, Alaska

Hughes DA, Münster F (2000) Hydrological information and techniques to support the determination of the water quantity component of the ecological reserve. Water Res Comm Report TT 137/00, 91 p. Pretoria, South Africa

Hughes DA, Hannart P (2003) A desktop model used to provide an initial estimate of the ecological in stream flow requirements of rivers in South Africa. J Hydrol 270:167–181

Iyer RR (2005) the notion of environmental flows: a caution. In: NIE/IWMI workshop on environmental flows, 23–24 March. New Delhi

Jones G, Hillman T, Kingsford R, McMahon T, Walker K, Arthington A, Whittington J, Cartwright S (2002) Independent report of the expert reference panel on environmental flows and water quality requirements for the River Murray system. In: Report prepared for the environmental flows and water quality objectives for the River Murray project board. URL: http://thelivingmurray2.mdbc.gov.au/__data/page/1482/ERPreport1.pdf. Accessed 30 Dec 2011

Jowett IG (1997) Instream flow methods: a comparison of approaches. Regulated Rivers Res Manage 13: 115–127

Karr JR, Dudley DR (1981) Ecological perspectives on water quality goals. Environ Manage 5:55–68

Land and Water Australia, Hamstead M (2007) Defining 'environmental flows', fact sheet for land and water Australia. http://lwa.gov.au/files/products/environmental-water-allocation/pf071349/pf071349.pdf

McLusky DS, Elliot M (2004) The estuarine ecology: ecology, threats and management. Oxford University Press, UK

Minshall GW (1984) Aquatic insect-substratum relationships. In: Resh VH, Rosenberg DM (eds) The ecology of aquatic insects. Praeger Publishers, New York, pp 358–400

N.C. Environmental Review Commission (2009) Report of the water allocation study. North Caroline Environmental Review Commission, North Carolina, USA. URL:http://sogweb.sog.unc.edu/Water/images/4/40/2008NCERCWaterAllocationStudyFinalReport.pdf. Accessed 03 Jan 2012

NCIWRD (1999) Integrated water resource development—a plan for action. National Council of Integrated Water Resource Development, Government of India, New Delhi

Petts GE (1996) Water allocation to protect river ecosystems. Regulated Rivers Res Manage 12:353–365

Silk N, McDonald J, Wigington R (2000) Turning in stream flow water rights upside down. Rivers 7 (4):298–313

Sinha RK, Prasad K (2005, Mar) Environmental flows vis-à-vis biodiversity: the current scenario in the River Ganga. Abstracts of the NIE/IWMI workshop on environmental flows. New Delhi, p. 13

Smakhtin VU, Revenga C, Döll P (2004) Taking into account environmental water requirements in global scale water resources assessments. Research report of the CGIAR comprehensive assessment programme of water use in agriculture. International Water Management Institute, Colombo, Sri Lanka, 24 p (IWMI Comprehensive Assessment Research Report 2)

Smakhtin V, Anputhas M (2006) Environmental flow requirements of Indian River Basins. Re-search Report No. 107. International Water Management Institute, Colombo

Smakhtin V et al (2007) Developing procedures for assessment of ecological status of Indian River Basins in the context of environmental water requirements. Research Report 114. International Water management Institute

References

Statzner B, Gore JA, Resh VH (1988) Hydraulic stream ecology: observed patterns and potential applications. J North Am Benthol Soc 7:307–360

Tare V et al (2015) Ganga River Basin management plan-2015. http://52.7.188.233/sites/default/files/GRBMP-MPD_March_2015.pdf

Ward JV, Stanford JA (1983) The intermediate-disturbance hypothesis: an explanation for biotic diversity patters in lotic ecosystems. In: Fotaine TD III, Bartell SM (eds) Dynamics of lotic ecosystems. Ann Arbor Press, Ann Arbor, Michigan, pp 347–356

WWF India (2012) Report on assessment of environmental flows for the Upper Ganga Basin. World Wildlife Fund India, New Delhi. http://awsassets.wwfindia.org/downloads/wwf_e_flows_report.pdf

Index

A
Accretion, 4
Active delta, 23
Adi Ganga, 88, 89
Ajay, 9
Alankhali, 66
Alluvial plains, 15
Amta channel, 102
Anderson Weir, 142
Atrayee, 27
Avulsion, 1, 22, 29, 53
Ayeen Akbery, 137

B
Baleswar, 50
Bangaduni, 50
Bank failure, 53
Bank protection, 68
Bansloi, 9
Barak, 9
Barind, 3, 15
Barrage, 36
Base flow, 13
Basin-fill history, 16, 19
Bay of Bengal, 3
Bengal basin, 1–3, 15, 16
Bengal Boundary Commission, 164
Bhagirathi, 77
Bhagirathi–Hugli, 9
Bhairab, 66
Bhairab–Jalangi, 50
Biodiversity, 13
Boral, 49
Braided channel, 31, 34, 73
Bund, 11
Buriganga, 74
Buriswar, 50

C
Canal mania, 140
Catchment, 100
Chandana, 68

Chandraketugarh, 88
Chars, 54
Chhoto Bhagirathi, 49
Chotanagpur plateau, 21
Churni, 50
Cloudbursts, 125
Coalescing delta, 23
Coastal erosion, 124
Coastline, 115
Colonial history, 2
Colonial hydrology, 11
Colonial legacy, 135
Colonial trade, 155
Community preparedness, 134
Creeks, 10
Cross-sectional area, 125
Cyclones, 119

D
Damodar, 9
Dampier-Hodges Line, 117
Dams, 146
Dams–canals network, 151
Darakeswar, 9
Debris-dam, 126
Deltaic plains, 23
Depocentres, 20
Depositional lobes, 27
Dhaka, 58
Disaster mitigation, 134
Discharge, 2, 98
Dispersal of organisms, 176
Drainage congestion, 96
Dredging, 139
Droughts, 133, 134
Duani barrage, 168
DVC, 13

E
Early warning, 134
Earthquakes, 133
Easterly flight, 22, 36

Eastward tilt, 80
Ecodeficit, 174
Ecohydrology, 171
Ecological flow, 166, 171
Ecological integrity, 174, 177
Ecologically critical area, 107
Ecological rapture, 3
Ecological services, 10, 13
Eden canal, 142
Embankment, 4, 12, 119
Environmental flows, 13
Erosion, 57
Extreme flood, 126

F
Family-level preparedness, 134
Fan, 3, 7
Farakka barrage, 49, 51
Farmlands, 138
Feeder canal, 83, 97
Flash flood, 125
Flood, 1
Floodable area, 126
Flood control, 146
Flood management, 2, 135
Floodplain, 84
Floodplain zoning, 135
Flood prone, 128
Flood-prone areas, 125
Fluvio-marine, 15, 20
Freedom from flood, 134
Freshwater, 119
Fulohar, 30
Funnel-shaped estuary, 9

G
Gajoldoba barrage, 161
Ganga–Brahmaputra (GB) delta, 15
Ganga–Brahmaputra–Meghna (GBM) delta, 1, 15
Ganga delta, 3, 15
Garai, 66, 68
Gaur, 51
Geometry of meander, 83
Gondwana coal, 146
Gosaba, 50
Groundwater table, 126

H
Haldi, 9
Haribhanga, 50
Helsinki Rules 1966, 164
Hijal Bill, 97
Hinge zone, 17
Holocene, 50, 80
Holocene delta, 19

Holocene period, 3
Homeless, 130
Horizontal connectivity, 173
Horizontal disconnectivity, 131
Hugli estuary, 91
Human intervention, 1, 176
Hydraulic structures, 128
Hydro-diplomacy, 63
Hydro-politics, 163

I
Ichhamati, 66
Indo-Bangladesh agreement, 166
Indo-Bangladesh border, 49
Intertidal zones, 111
Irrigation, 13
IUCN, 179

J
Jaldhaka, 27
Jamuna, 73

K
Kalindri, 49
Kana Damodar, 141
Kana Nadi, 141
Kananadi, 138
Kansai, 9
Kanshabati barrage, 152
Karatoya, 27
Kata Ganga, 89
Kobadak, 66
Kolkata, 58
Kolkata Port Trust, 9
Konar, 147
Kopai, 98
Kosi, 52
Kulik, 27
Kusiyara, 74

L
Landslides, 126
Largest delta, 1
Lateral oscillation, 59
Lateritic tracts, 15
Littoral tract, 3, 107
Longitudinal disconnectivity, 1, 173

M
Madhupur, 15
Mahananda, 22, 27
Malancha, 50
Malaria, 11

Index

Malarial epidemics, 11
Mangrove, 109
Mathabhanga, 50
Mathabhanga–Churni, 77
Matla, 50, 115
Mature delta, 23
Mayurakshi, 9
Mayurakshi project, 152
Meandering, 40
Meander loop, 83
Meander migration, 1, 52
Mega fan, 34
Meghalaya plateau, 15
Meghna, 50
Meghna estuary, 1
Moja Damodar, 141
Moribund delta, 23
Mudflats, 111
Mundeswari, 102
Muriganga, 50
Murshidabad, 58

N

Navigation, 2, 12
Navigational route, 81
Navigation channel, 91
Neo-refugees, 68
Neo-tectonism, 80
Nodal point, 83
Non-tidal regime, 3

O

Off-take, 82
Overflow irrigation, 130

P

Pagla, 9, 49
Palaeo-channels, 4
Palaeo-strandlines, 17
Panchet, 147
Pandua, 51
Patuakhali, 50
Photosynthetic exchange, 175
Piedmont Zone, 29
Plate tectonics, 15
Pleistocene deposit, 22
Pleistocene low stand, 20
PMUD, 145
Polders, 4, 119
Pollen analysis, 109
Port of Kolkata, 83, 89
Porto pequeno, 87
Premature reclamation, 4, 12
Punarbhava, 27
Pusur, 50

R

Rajmahal–Meghalaya gap, 27, 125
Rarh Bengal, 4, 12
Reclamation, 118
Rejuvenation, 67
Riverine ecosystems, 174
River management, 12
RMSL, 121
Rupnarayan, 9

S

Sankosh, 31
Saptamukhi, 115
Saraswati, 86
Satgaon, 58, 87, 153
Sea level, 18
Sea-level change, 121
Sea surface temperature, 121
Sediment dispersal, 2, 4, 125
Sediment influx, 3
Sediment load, 20, 24
Selagang, 50
Self-recovery, 176
Sepoy Mutiny, 140
Shelving bank, 64
Silai, 9
Silt-laden, 4
Skewed rainfall, 138
Sorrow of Bengal, 130
South-west monsoon, 20
Submarine delta, 16
Subsidence, 29, 116, 121
Sundarban, 3, 12
Surma, 74
Suspended load, 63
Swatch of No Ground, 18
Sylhet basin, 3, 15, 19
Sylhet trough, 74

T

Tamralipta, 79, 87
Tanda, 51
Tarai–Doors, 27
TBP, 157
Tectonic unit, 1, 3
Teesta, 22, 27
Teesta barrage Project, 13
Teesta fan, 29
Teesta–Jaldhaka Main Canal, 159
Teesta–Mahananda Link Canal, 158
Tenughat, 148
Thakuran, 50, 115
The red-water famine, 139
Tidal creeks, 4
Tidal surges, 119, 126
Tidal upsurge, 121

Tide-velocity asymmetry, 112
Tilaiya, 147
Tilpara barrage, 152
Tolly's Nala, 117
Torsa, 27
Transboundary rivers, 30, 163
Tropical cyclones, 120

U
Unrestricted flow, 174

Z
Zamindars, 139

Printed by Printforce, the Netherlands